Collins Colour Guides

MINERALS
AND PRECIOUS
STONES

COLLINS COLOUR GUIDES

MINERALS AND PRECIOUS STONES

RUDOLF METZ

Translated from the German by

G. A. WELLS

Illustrated with 150 colour photographs

A. E. FANCK

COLLINS
St. James's Place, London

The picture of manganese spar on the jacket of this book is an enlargement of Plate 57. First published in German by Chr. Belser Verlag, Stuttgart, under the title EDLE STEINE, in 1965.

ISBN 0.00 212110-7

FOREWORD

The book by R. METZ and A. E. FANCK entitled *Antlitz edler Steine* has hitherto been available only in a somewhat expensive edition in large format. The present volume extends the text of this work, and presents it in a smaller format to bring it within the means of the average mineral collector and amateur, and the professional geologist can only greet the work with unqualified approval. The dramatic colour plates will assist and stimulate the amateur in his observation and analysis of the real material. He will not normally be the owner of a precious specimen such as the proustite of Chanarcillo, but this book will teach him how to classify and care for such stones as he does have.

The mineralogist welcomes this book on another ground too. It is mainly books like this which lead young people to take an interest in science, and some of them may in consequence be influenced in their choice of career. In science young people are needed who not only have a thorough grasp of their subject, and are impelled by a sense of duty to work hard and publish their findings; they must also feel real enthusiasm, and in mineralogy this means being fascinated by crystals, minerals, rocks and the secrets they enclose. It is my conviction that this book will do much to arouse such feelings.

PAUL RAMDOHR

LIST OF CONTENTS

INTRODUCTION

During the course of history radical changes have occurred in the values put upon individual minerals; but since the Stone Age the significance of mineral raw materials for everyday life has increased continually, and bitter conflicts have been waged to secure the localities where important minerals occur as ores.

There is something enigmatic about crystal forms that has always appealed to man's imagination. Mythology is full of precious stones possessing magic powers, conferring magic capacities to their owners — and the religious and superstitious elements of these traditions survive to some extent even today. What it is that makes minerals, and particularly mineral assemblages, beautiful is hard to put into words: it is perhaps a combination of the great variety of crystal forms, their properties of symmetry, their colours, and the various effects due to reflected and transmitted light.

Dr. Ernst Petersen, who died in 1959, planned to show the characteristics and richness of minerals in a volume of pictures, and he found in Arnold E. Fanck an exceptionally patient and skilful photographer.

Thanks are especially due to those who made available minerals in the collections and museums entrusted to their care, and who helped in many other respects as well — in particular DR. E. JÖRG in Carlsruhe, DR. O. GRÜTTER in Basel, DR. M. SCHNETTER in Freiburg (Breisgau), DR. M. GRÜNENFELDER in Zürich, DR. H. DACHS in Munich, and Professor H. WONDRATSCHEK of Carlsruhe. Herr W. FINCK in Freiburg was kind enough to allow minerals in his private collection and from his firm to be photographed, and Herr A. PAN-

ZER of Oberrotweil (Kaiserstuhl) made assemblages from his collection available.

The Plates in this volume have been selected from a large number. After photos which aim at showing the general properties of minerals there follow some examples of the distinctive features of a number of mineral parageneses. This book does not set out to be a textbook, but tries to show minerals as they occur in nature, in the form and colours proper to them. For this reason considerable emphasis is placed in the text on those mineral parageneses that

◁ 1   AMETHYST a violet variety of the mineral quartz. A group of prismatic crystals with terminal pyramid faces. The colouration is caused by a small content of manganese and titanium. From Schemnitz, Erzgebirge, Czechoslovakia. Greek *amethystos,* counteracting intoxication, since the mineral was valued as an amulet against drunkenness. SiO_2 trigonal. Scale 3:1

◁ 2   LABRADORITE a dark violet-grey plagioclase (calc-soda felspar); coarse fragment bounded by fracture planes. The angular area on the right shows the characteristic iridescence of labradorite (labradorescence). From St. Paul's Rocks, islets in the Atlantic. Named from its occurrence on the coast of Labrador. Mixed crystal of albite (Latin *albus,* white; $Na[AlSi_3O_8]$), and anorthite (Greek *anorthos,* askew, because of its inclined crystal faces; $Ca[Al_2Si_2O_8]$), both triclinic. Scale 2:1

◁ 3   GALENA in lead-grey, cube-shaped crystals; on them lie reddish zincblende and yellowish calcite. From Joplin, Missouri, U.S.A. Galena PbS cubic. Zincblende ZnS cubic. Calcite $CaCO_3$ trigonal. Scale 1·5:1

are included in the illustrations. For fuller treatment of minerals, their properties, occurrence and uses, the reader is referred to the select bibliography on p. 255.

TRANSLATOR'S NOTE. The translator is glad to record his thanks to DR. M. K. WELLS (Reader in Mineralogy in the University of London) for advice offered on a number of points.

MINERALS
CHARACTERISTICS AND DISTRIBUTION

1   MINERALS AND CRYSTALS

MINERALS AND CRYSTALS. The word *mineral* comes from the Latin *minare*, to mine, and originally designated minerals and rocks so obtained. Today we understand by minerals the naturally occurring components of the earth's crust which are of uniform composition. Usually several species go to make a rock. Only a few rock types are monomineralic, formed from a single mineral species, like for example marble consisting only of calcite grains.

The name *crystal* comes from the transparent rock crystal which the Greeks called *krystallos,* ice (Pl. 102). In antiquity and even in the early Middle Ages it was thought that this mineral rock crystal was simply ice that had been frozen so hard by exposure on high mountains that it could not be thawed. Then 'crystals' came to have the wider meaning of naturally occurring minerals bounded by plane faces and straight edges. However, artificial products such as metals or salts can likewise crystallize with plane faces.

The expression *crystal form* designates a specific ordered state of matter. Crystalline materials have a lattice structure on an atomic scale. This structure is not only expressed in the capacity to assume crystal form; it has the further effect that many physical properties in such bodies vary according to the direction in which they are measured. In consequence crystals, as a special state of solid matter, are in a class by themselves distinct from gases

and liquids. In many cases the capacity of a material to build regular crystalline faces may not be apparent: the outer configuration is arbitrary and contingent because free growth was prevented. Today those crystalline substances are called crystals (although this usage is not uniform) which can at least to some extent develop their own crystalline form. There are only a few amorphous minerals, such as opal, hyalite or amber. Amorphous minerals are never bounded by plane faces; they are never granular or fibrous in texture: instead they commonly have rounded surfaces which look like droplets (Pl. 20). Amorphous minerals can in time become crystalline. Thus as a result of internal changes opal can become chalcedony.

Merely by looking at a well-developed crystal with many faces one senses unconsciously the laws of *crystallographic symmetry*

◁ 4   RHODOCHROSITE red, spherical aggregates resembling raspberries, on encrusting honeycombed limonite. From Wolf mine, Herdorf, near Betzdorf in valley of the Sieg. Rhodochrosite (Greek *rhodochroos,* rose-coloured) $MnCO_3$ trigonal. Scale 5·1:1

5   SULPHUR crystals of sphenoidal habit, resting on a surface of ▷ calcite. The smaller, nearer sulphur crystal has a broken point. From Ciaciana, Sicily. Sulphur S orthorhombic. Scale 1·8:1

6   BARYTES (heavy spar) blue, translucent, tabular crystals. On ▷ the right, brown-violet haematite as a layered nodule and yellow, younger barytes in small crystals. From Cumberland. Barytes $BaSO_4$ orthorhombic. Haematite Fe_2O_3 trigonal. Scale 3:1

which put minerals in a different class from organisms. A further significant difference between minerals and the organic world is the size of individuals. Fairly rigid ideas of magnitude are linked with the names of particular animals or plants, but the size which minerals can attain is almost unlimited and depends only on whether the conditions of formation persist long enough, whether material is supplied without interruption to the growing crystal, and on sufficient space being available. There are crystals which measure only a fraction of a millimetre, and others of the same mineral species which weigh over 100 tons.

The NUMBER OF MINERALS is surprisingly small. About 3,000 species are known today. It is true that a few new ones and modifications of existing ones are added every year, but on the other hand some cease to be listed as independent minerals because they have proved to be mixtures or very fine intergrowths of others. Hence the number of known mineral species only increases very slowly. Compared with, for instance, the number of chemical compounds — organic chemistry alone recognises nearly half a million carbon compounds — the number of minerals formed as naturally-

7 CALCITE yellowish crystals of squat habit with rough faces; ▷ small white rhombs of dolomite, brownish rhombohedral crystals of brownspar. Below: white and reddish coloured compact barytes. From Sophia mine, Wittichen, Black Forest. Calcite from Latin *calx*, lime. $CaCO_3$ trigonal. Dolomite after the French mineralogist Dolomieu. $CaMg[CO_3]_2$ trigonal. Brownspar $(Mg, Fe, Mn) CO_3$ trigonal. Barytes $BaSO_4$ orthorhombic. Scale 2·3:1

occurring combinations of elements is amazingly small. Mineral species also compare very modestly in number with botanical or zoological species. The number of known flowering plants or insects is vastly greater.

The idea of *mineral species* is borrowed from the organic kingdom, where the smallest subdivision that can interbreed is called a species. But since this criterion is inapplicable to minerals, the species boundaries — particularly within the widely distributed mixed-crystal series – are often quite arbitrary. It is likewise following the terminology of biology that the single crystal is called an individual.

THE FREQUENCY OF MINERALS. Even more astounding than the relatively small number of mineral species is the very different extent to which they contribute to the accessible part of the earth's crust. More than half of this is formed by felspars. After quite some interval there follow, as the next most frequently occurring minerals, the pyroxenes and amphiboles, represented by their most important members augite and hornblende. Next come quartz and mica (dark biotite and light muscovite). These few mineral groups already comprise well over 90%, and the modest remainder is shared between olivine and the widely distributed accessory minerals of magmatic rocks, e.g. magnetite, zircon, ilmenite and rutile. We must add to these the minerals important in metamorphic rocks, namely chlorite, serpentine and garnets, and the clay minerals and carbonates (especially calcite and dolomite) of sediments. Finally, among the more common minerals are haematite, pyrite, limonite, the felspathoids, titanite, chromite, corundum, tourmaline, the spinels, pyrite and chalcopyrite. The few mineral species named here constitute more than 99% of the solid crust, and all others, including the beautifully crystalline ore-minerals — the pride of every collection — and the much-coveted precious stones, form quantitatively a quite insignificant portion of it. Since the crust consists for the most part of magmatic and highly metamorphosed rocks that are akin to magmatic ones, the predominance of magmatic mineral species is understandable.

MINERALS AND CHEMICAL ELEMENTS. The most common element in the solid crust is oxygen, which accounts for almost half of the crust by weight. More than a quarter consists of silicon, and then follow aluminium, iron, calcium, sodium, potassium, magnesium and titanium. The crust is essentially built of these nine elements, and all other chemical elements form just 1% of its weight.

From the frequency with which the various elements occur it will be obvious that silicates and quartz are the minerals most widely distributed. Nevertheless the frequency of the elements is by no means reflected in the number of the minerals in which they form essential components. Thus there are elements which are still relatively common, but which do not enter decisively into the composition of any mineral — for example rubidium, which holds seventeenth place in frequency of occurrence, but occurs in any quantity only in the rare pegmatite mineral rhodosite. Hafnium too, which is far commoner than antimony or bismuth, does not go to make any special mineral. On the other hand silver, forming only 0·00001% of the crust by weight, and coming almost last among the elements, forms an appreciable number of independent mineral species. The example of galena shows that a relatively rare element like lead can nevertheless form a common mineral; lead is present in the earth's crust only as 0·003% by weight.

This capriciousness in the behaviour of the elements and their different capacity to form distinct minerals is founded in their geochemical properties. Many elements are characteristically associ-

CONSTANCY OF ANGLES. As a crystal grows, the size of adjacent crystal faces naturally changes; but the angle which the two faces contain remains constant. Different sized crystals of the same mineral species are therefore similar, provided they have the same crystal form. Each mineral has quite definite angles (which are often diagnostic of it) between the faces belonging to the same form. But the surface angles are not only characteristic in the case of crystals that have grown naturally; they appear also in artificially produced crystals of the same substance. And finally, the angles of cleavage fragments obey the same crystallographic laws. Thus the cleavage angle in calcite rhombs is always 105° 5′, no matter from what locality the fragment has come, nor from what paragenesis. It is therefore possible by means of these interfacial angles to determine minerals and to distinguish them from other kinds.

ELEMENTS OF SYMMETRY. Crystal faces are regularly distributed, and this regularity is known as symmetry. In the many symmetry properties of crystals the atomic structural principle which governs 'dead matter' is particularly clearly expressed. If a crystal can be divided by a given plane into two halves such that the one is a mirror image of the other, then this plane is called a *mirror plane* or *plane of symmetry*. A crystal is said to possess a *centre of symmetry* if every one of its faces has a complementary face in mirror image relationship. A third kind of symmetry elements are axes, around which one can imagine the crystal to be turned until it reaches a position identical with the one from which it started. According to how often a crystal reaches parity with its initial position in the course of one complete revolution round such an *axis of symmetry*, we can distinguish twofold, threefold, fourfold and sixfold axes of symmetry. The angle of rotation round the axis amounts therefore in each case to 180°, 120°, 90° or 60°,

ated with others, forming as it were impurities to them. Thus the zircons widely distributed in magmatic rocks usually contain some hafnium and thorium. Such elements are therefore present 'in disguise' in other minerals and were in consequence in many instances long overlooked. Conversely there are rare elements which for crystallographic and chemical reasons nevertheless enter into a surprisingly large number of distinct mineral species. For instance the large variety of antimony ores is out of all proportion to the small part which the elements composing them play in the constitution of the earth's crust.

MINERAL NAMES. The history of mineralogy is reflected in the multifarious confusion of mineral names. Many terms have reached us from antiquity, or from the east, others derive from mediaeval German mining usage, e.g. glance-ores, sparstones. Other names originate from superstition or mythology, or indicate the country of origin or the locality where the mineral was found. Chemical composition, appearance or other distinctive properties have like-

◁ 8   FLUORSPAR (fluorite) grey cubes with violet or brown-coloured crystal edges. From Cecilia mine, Wölsendorf, Upper Palatinate, Bavaria. Fluorite from Latin *fluor*, liquid, because used a flux in metallurgy. CaF_2 cubic. Scale 1·5:1

9   MALACHITE polished surfaces of concentrically layered nodules. ▷
10 Between the individual nodules there are in places thin crusts of limonite and haematite. From Gumishev in the Urals, USSR. Greek *malache*, mallow, because of its colour. $Cu_2[(OH)_2/CO_3]$ monoclinic. Scale 1·6:1

corresponding to the rectangular, triangular, quadrate or hexagonal form of the elementary cells, the smallest units of construction of the crystals. In the same way, a surface, e.g. a floor, can be completely covered with equal sized bricks only if they are rectangular (symmetry number = 2), equilateral triangular (3), square (4) or regularly hexagonal (6). It cannot be done if the units have five, seven or eight equal sides, and so the world of crystals is lacking in a fivefold, sevenfold or eightfold symmetry, such as are found in organisms like radiolaria, corals, molluscs, starfish, or in flowering plants.

The higher the symmetry according to which a crystal is constructed, the greater the number of axes and planes of symmetry that can be distinguished in it. Crystals range from the triclinic forms of lowest symmetry, which have no symmetrical properties whatever, to the highly symmetrical cubic forms having 9 planes, 3 fourfold, 4 threefold and 6 twofold axes, and a centre of symmetry. The symmetrical properties of crystals are also expressed in the fact that faces in symmetrical relationship have the same properties. Thus in many crystals rough and matt faces can be seen alongside lustrous and reflecting ones, or there is an alternation between completely smooth and striated faces. If striated or grooved faces occur, the striae follow the symmetry of the mineral's crystal form (Pl. 13). The same is true of foreign particles which are embedded in regular orientation.

CLASSES OF SYMMETRY AND CRYSTAL SYSTEMS. The numerous possibilities of crystal form can be reduced to 32 classes of symmetry, divided between 7 crystal systems, namely the cubic (or regular), the tetragonal, the hexagonal, the trigonal (or rhombohedral), the rhombic (or orthorhombic), the monoclinic and the triclinic. The commonest minerals of the earth's crust, the felspars, crystallize in the monoclinic and triclinic systems.

ATOMIC STRUCTURE. Individual mineral species are characterised not only by their composition but also by a definite arrangement of the smallest units of construction. The regular arrangement of the smallest particles (atoms, ions, molecules) in space is called lattice structure, or commonly atomic structure. The smallest grouping of the individual units which by endless repetition make a crystal is called an *elementary cell* or *unit cell*. A space lattice of this kind on an atomic scale is present in all crystalline substances. By combining the simplest space lattices (*elementary lattices*) according to the possibilities of symmetry, one obtains 230 different *space groups,* which comprise the whole variety of all crystal lattices. Sodium chloride, rock salt, which crystallizes in cubes, has one of the simplest lattices of all minerals, each chlorine ion being

◁ 12 PSILOMELANE encrusting, stalactiform. From Eiserfeld, Hessen. Psilomelane from Greek *psilos,* bald and *melas,* black. Predominantly MnO_2 orthorhombic. Scale 1·6:1

13 CALCITE colourless transparent crystals of prismatic habit. ▷ Striations corresponding to the trigonal symmetry on the terminal faces. From Gillfoot mine, Egremont, Cumberland. Latin *calx,* lime. $CaCO_3$ trigonal. Scale 1·3:1

14 CALCITE aggregates in wheat-sheaf form; prismatic white crys- ▷ tals with brown terminations. From Crucimesti-Boita in Transylvanian Erzgebirge. $CaCO_3$ trigonal. Scale 1·3:1

15 ARAGONITE prismatic, chisel-shaped transparent crystals. From ▷ Bilin, Bohemia, Czechoslovakia. Named after Aragon, Spain. $CaCO_3$ orthorhombic. Scale 2·3:1

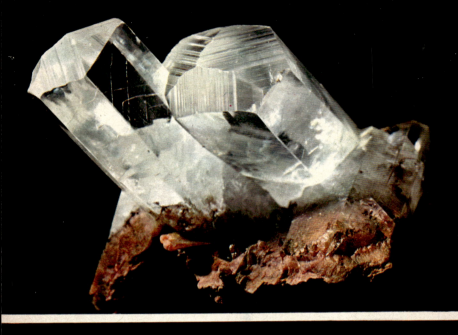

surrounded by six neighbouring sodium ions, and each sodium ion likewise having six neighbouring chlorines (Pl. 16). The distance from the centre of one ion the nearest like ion is $5·64 \times 10^{-8}$ cm or 5·64 ten millionths of a millimetre. Thus on the edge of a cube of rock salt a cubic millimetre in size there are nearly 2 million elementary cells alongside each other. The determination of the atomic structure of minerals from X-rays forms an important branch of crystallography.

MINERAL SPECIES, VARIETY, MODIFICATION. Many chemical substances can crystallize in a number of different atomic structures, thus forming several naturally occurring modifications and so several mineral species. Calcium carbonate, for instance, occurs in the two crystal forms of calcite and aragonite (Pls. 13, 15). Among the dimorphous substances, which can form two distinct mineral species, is also iron sulphide, which we meet as pyrite and as marcasite (Pl. 60). Titanium dioxide even forms three minerals and is therefore said to have three *polymorphs,* namely rutile, anatase and brookite. Also trimorphous is an aluminium silicate which gives andalusite (Pl. 86–7), kyanite and sillimanite. In all these examples the mineral species is determined by the chemical composition and a particular lattice structure.

16 ROCK SALT (halite) cubic crystals grown in clay from the so- ▷ called 'Haselgebirge' — an Austrian mining term for rocks in which salt crystals or crystal aggregates seem to swim in clay. From Hall salt works, Tirol. In the background a model of the salt lattice. The red and blue spheres mark the positions of the sodium and chlorine ions. Halite from Greek *hal*, salt. NaCl cubic. Scale 1·3:1

In many crystal lattices individual particles can be replaced by others of a different kind, provided they are of similar size and have the same electric charge (*diadochy* or equivalent atom substitution). When such an exchange of units occurs, the lattice structure is in essence preserved, but the chemical composition is changed. In such cases one speaks of *mixed-crystal series.* Among minerals, mixed crystals with continuous transitions are of frequent occurrence, and as a result the boundaries between mineral species have to be defined arbitrarily. In a mixed-crystal series important members are often separately named. Olivine (Pl. 23), important as a rock-forming mineral, occupies a place in a series between a magnesium-rich (forsterite) and an iron-rich (fayalite) end-member. The calc-soda felspars (plagioclases), forming more than a third of the solid crust, form a continuous series between a sodium-rich (albite) and a calcium-rich (anorthite, Pl. 2) end-member. If a mineral species sometimes has individuals with abnormal *colouration* or special *crystal forms,* these varieties are often given special names. Thus amethyst is quartz coloured violet by small quantities of iron or manganese (Pl. 1); and the transparent variety of beryl coloured by traces of chromium is known as emerald (Pl. 35). Superfluous and often even misleading designations of colour variants are particularly common for minerals used as precious stones and jewels.

Many substances can form different minerals under different environmental conditions. Potash felspar appears as microcline in many deep-seated rocks, as amazonstone in pegmatites, as orthoclase (Pl. 19) in cavities in granites (in the so called vugs), as sanidine in volcanic rocks, and finally as adularia (Pl. 101) in alpine joints.

CRYSTAL GROWTH. Crystals grow from *centres of crystallization* by accumulation of homogeneous material on the surface of a crystal nucleus — as opposed to organisms, whose growth is conditioned by metabolism. Material can be deposited at the same rate on all the faces of a growing crystal, so that they retain their shape as they increase in size. But if certain faces are preferred to others, the crystal changes its configuration, its habit, as it grows, and assumes another crystal shape. If a substance causing colouration or opacity lies on certain faces of a growing crystal, a zoned structure can result which allows earlier stages of growth to be recognised (Pls. 8, 97). Normally there is a slow and steady change in the composition, concentration, pressure and temperature of the solution in which the crystals grow; and any materials present to cause colouration or opacity increase or decrease at a similar slow, steady rate. In this way there arise long columnar crystals which are coloured differently or of different transparency at their ends. Prismatic rod-like minerals which have been formed in vugs are often opaque at their roots of attachment and gradually pass up into perfectly transparent or purely coloured crystal terminations (Pls. 1, 102). If the supply of liquid is interrupted crystal growth is of course brought to a halt. In the case of shallow, disc-like aggregates (Pl. 70-1) all the numerous needle-like crystals ceased growing simultaneously.

RICHNESS OF FORM. The degree of complexity of combinations of crystal forms which can occur varies considerably between individual mineral species. Some minerals are not known to develop crystals with terminal faces (e.g. the germanium ore germanite);

many can only rarely form well developed crystals, or are only capable of simple forms. But others, like topaz, garnet, potash felspar, galena, pyrite or chalcopyrite, have a great variety of forms. This is particularly true of those of hydrothermal origin, i.e. those precipitated from aqueous solutions at high temperature. Thus more than 200 different crystal forms of barytes are known; and the mineral with greatest variety of form is calcite, with several hundred possible positions for the faces, and combinations

◁ 17 CALCITE double refraction of water-clear crystal — a cleavage rhomb of Iceland Spar. The coloured strip appears displaced and doubled through the crystal. (The edges of the doubled strip overlap to give a dark central line.) From Helgustadir, Eskifjord, Iceland. $CaCO_3$ trigonal. Scale 1·2:1

18 LABRADORITE (plagioclase). Grey fragment with metallic-blue ▷ light effects (labradorescence). The surface is a fracture plane. Twin lamellae can be recognised. The specimen is from the type locality of Labrador, Canada. Mixed crystal of albite, $Na[AlSi_3O_8]$, and anorthite, $Ca[Al_2Si_2O_8]$; both triclinic. Scale 1:1

19 TOURMALINE group of three crystals with quartz and orthoclase. ▷ The largest elongated prismatic tourmaline has longitudinal striae and is not uniformly coloured: green at the bottom, rose in the middle, almost colourless at the top. From San Pietro Campo, Island of Elba. Tourmaline after the Singhalese term *turmali*. $(Na, Ca) (Li, Mn, Mg, Fe, Al)_3 Al_6 [(OH, F)_4/(BO_3)_3/ (Si_6O_{18})]$ trigonal. Orthoclase from Greek *orthos*, right-angled and *klao*, cleave. $K[AlSi_3O_8]$ monoclinic. Scale 2·3:1

of these faces giving more than a thousand forms. Often a mineral at a single ore horizon occurs in a number of crystal forms. 144 simple forms and 391 combinations of forms of calcite are known from the ore veins at the single locality of St. Andreasberg in the Harz.

The richness of many crystal species in the forms which they display is conditioned by environmental factors. Just as an organism is moulded by heredity and environment, a crystal is decisively influenced in the forms it assumes not only by its lattice type, but also by many factors in its environment: temperature, pressure, concentration, other ions in the solution, and also the rate of crystallization — that is to say time, which is available to nature in rich measure.

IDEAL CRYSTALS AND REAL CRYSTALS. Most naturally-occurring crystals are imperfectly developed, because they are sited on some foundation, such as the wall of a vug. Minerals suspended in a plastic medium are well developed in all directions, e.g. gypsum crystals in marl or clay. Irregularity in the supply of material during growth gives rises to distortion. Growth can also be obstructed by adjacent rocks or nearby minerals of another species, and as a result of such external factors natural crystals often deviate markedly from their *ideal form*. Apart from coarser defects, impurities, distortions and defects in growth, there are also disturbances and mal-alignments of the order of magnitude of the lattice units, such as incorporated foreign particles. Crystals of any

20 OPAL encrusted on sandstone with blue-white play of colours ▷ migrating over the surface (opalescence). From Adamooka, S. Australia. Opal from Greek-Latin *opalus* from Sanscrit *upala*, gem-stone. $SiO_2 \cdot nH_2O$ amorphous. Scale 1·7:1

appreciable size are therefore never of perfectly uniform construction, but consist of mosaics of lattice blocks slightly displaced relatively to each other. A natural crystal with its distortion, impurities and lattice defects is a *real crystal,* as against the mathematical ideal lattice model, the *ideal crystal.*

IDIOMORPHS AND XENOMORPHS. Minerals which have been able to develop the crystal form proper to them are called *idiomorphic* (having their own form). Minerals formed between other already existing crystals and whose bounding faces are therefore contingent are called *xenomorphic* or *allotriomorphic* (having a foreign form). Completely idiomorphic crystals, perfectly developed in all directions, are uncommon. Crystals usually grow on some basement material and are therefore idiomorphic only in certain directions (Pls. 1, 15, 19, 102).

GUISE AND HABIT. The sum total of faces bounding a crystal is known as its *guise,* while the general configuration of a crystal is its *habit,* and may be acicular, rod-like, columnar, prismatic or tabular. Just as two men may be of fat or thin 'habit' and yet may both wear the same clothes, so can, say, two barytes crystals be of thick or thin tabular habit, and nevertheless have exactly the same faces, i.e. the same guise. On the other hand minerals can have the same habit but different guise, as when two mineral species both form crystals of prismatic habit, although the crystals of the one exist in hexagonal, and those of the other in tetragonal guise. In many cases the conditions under which crystals were formed can be inferred from their guise and habit, and sometimes these characters are completely diagnostic of a mineral's mode of origin.

INTERGROWTHS. Minerals are often intergrown to form aggregates, whether or not they are of the same species. Most mineral collections exhibit not only individual crystals, but such associated aggregates which comprise identical or discrete mineral species, and which have a far greater appeal than individual specimens (Pls. 3, 6, 7, 43, 53–4, 78–9, 91).

Minerals often consist of chance and random intergrowths, but there are also crystals which are regularly intergrown, and these are of particular interest. Such *epitaxic* or orientated growths of one mineral species upon another are known from numerous examples. They depend on some shared characteristic in the atomic structure of the host mineral and the mineral intergrown with it.

TWINS. A notable property of many mineral species is their capacity to form twins, i.e. regular symmetrical intergrowths of two crystals of the same mineral species. If more than two crystals are linked in a regular relationship we speak of doublets, triplets or 'multiple twins'. Twins arise in the very first stages of growth as a result of a regular relationship between centres of crystallization.

Ability to form twins is an atomic lattice property and varies considerably from one mineral species to another. While many minerals, e.g. andalusite, are not known to form twins at all, others form many kinds of them. Felspar, for instance, forms twins according to laws known as Carlsbad, Baveno and Manebach, and rock crystal according to the Brazilian, Japanese, Dauphiné and Swiss laws, as well as according to less common ones. Many minerals can easily be recognised from their twins. Thus staurolite forms characteristic skew and right-angled interpenetrant twins (Pls. 95, 96). A significant character from which twinned intergrowths can be recognised are re-entrant angles, which cannot occur in single crystals, for crystals whose growth is unimpeded always occupy space by assuming a convex form (Pl. 103).

DISTORTED CRYSTALS. As a result of irregularities in the supply of materials in solution, caused e.g. by currents in the mother liquor,

growing crystals can assume distorted forms, i.e. minerals in which some faces are greatly enlarged, while others, their symmetrical equivalents, are suppressed. Under such conditions fluorspar gives rise not to normal cubes but to prismatically elongated individuals (Pl. 53–4). With native metals distorted crystals are very common (Pl. 40). As a sport of nature, distorted crystals — and twins too — mimic a symmetry alien to the relevant minerals, and in such cases of pseudosymmetry (Pl. 25) the underlying ideal crystal form is barely discernible.

SKELETAL CRYSTALS. When the terminal points and edges of a crystal grow precipitately and outstrip the growth of the faces, skeletal and mesh-like configurations may be formed. Most snow flakes belong in this class, as do also the 'window-quartzes' (Pl. 98). The so called dendrites — brown or black skeletal crystal aggregates with patterns of ice-flowers, ferns, mosses or trees inscribed on them — were formed by iron or manganese solutions which penetrated along bedding planes or into joints and crystallized out when the solvent dried up (Pls. 62, 63).

◁ 21 MUSCOVITE tabular brown crystal showing cleavage resembling the leaves of a book. From Dalby, Sweden. Muscovite from Latin *Muscovia*, Moscow, since transparent mica-plates were imported from Russia. $KAl_2[(OH, F)_2/AlSi_3O_{10}]$ monoclinic. Scale 1·5:1

◁ 22 COVELLITE (copper indigo), leaf-like thinly tabular crystals with bloomed colours. The iridescence is caused by very thin oxidation films and the high dispersion of covellite. From Alghero Sassari, Sardinia. Covellite after the Italian mineralogist Covelli who discovered it. CuS hexagonal. Scale 1·8:1

MINERALS FORMED FROM GELS. Nodular or reniform mineral aggregates with smooth, lustrous surfaces have mostly originated from an aqueous gel of loamy consistency. Such gels consist of very small particles of mineral substance. On drying, the slimy crusts and nodules gradually harden; they often assume a crystalline structure on an atomic scale, and finally become quite solid. An old mining term for such round shining bodies is 'glebe'.

Few minerals (and only relatively young ones) which were formed in a gelatinous state are still amorphous today, like opal (Pl. 20). Gel minerals that have in the course of time become crystalline

◁ 23 OLIVINE a broken 'olivine bomb', with an exterior dark violet-grey scoriaceous crust. The inside of this product of volcanic ejection consists of an aggregate of olive-green olivine grains. Between them are isolated crystals of emerald green chrome-diopside. From the lake near Dreis, Eifel. Olivine from its olive-green colour. $(Mg, Fe)_2 [SiO_4]$ orthorhombic. Chrome-diopside from Greek *chroma*, colour; *dis,* twofold; *opsis,* appearance, because of its crystal form. $CaMg[Si_2O_6]$ monoclinic. Scale 2·6:1

◁ 24 AUGITE a phenocryst weathered from a volcanic eruptive rock. From Mondhalde, near Oberbergen, Kaiserstuhl. Greek *auge,* lustre. $(Na, Ca) (Mg, Fe, Al, Ti) [(Al, Si)_2O_6]$ monoclinic. Scale 5·5:1

25 BIOTITE 'books' of thinly tabular pseudohexagonal crystals in ▷ a vesicle of a volcanically ejected rock. From Vesuvius, Italy. Biotite after the French physicist Biot. $K(Mg, Fe, Mn)_3 [(OH, F)_2/(AlSi_3O_{10})]$ monoclinic. Scale 2·3:1

can be recognised by their beaded surface, covered with black-berry-like spheres, and their internal texture of concentric layers or radial fibres, i.e. by the arrangement of the individual crystals. In addition to spherical, warty forms (Pl. 69), minerals formed from gelatinous masses are often strobiliform or stalactitic in appearance.

MINERAL AGGREGATES. Sizable individual crystals with terminal faces perfectly developed all round are rare in nature. Usually numerous crystals all grow at the same rate and interfere with each other's growth. Accumulations of identical minerals, built of numerous closely packed individuals, are called *mineral aggregates*. According to their structure and appearance they are called radial-fibrous or radial acicular (natrolite Pl. 28; wavellite Pl. 70–1), concentrically layered (malachite Pl. 9–10), irregular granular (lapis lazuli Pl. 90), botryoidal, reniform (blende Pl. 51; azurite Pl. 75), nodular (smithsonite Pl. 83), encrusting (antimony ochre Pl. 85), strobiliform, warty (chalcedony Pl. 30; chrysocolla Pl. 69; adamite Pl. 84), stalactitic (limonite Pl. 65), fibrous (malachite Pl. 73), fascicular, wheat-sheaf shaped (calcite Pl. 14), scaly (chlo-rite Pls. 101, 103) or crazy fibrous (amianthus Pl. 105) — the abundance of forms assumed by mineral aggregates is almost in-exhaustible. Joint planes are often covered with a matted crystal coating of numerous close-set small minerals. Native metals such as copper, silver or gold often occur as wire or foil, or in mossy or dendritic form (Pls. 11, 45, 46).

INCLUSIONS. Many crystals have incorporated foreign particles during growth, and in consequence become opaque or of different appearance. Sometimes gaseous or liquid residues of the mother liquor are engulfed. Quartz often acquires a milky turbidity from inclusions of extremely fine liquid particles invisible to the naked eye (milky quartz). Larger and quite distinct crystals of chlorite, amianthus, tourmaline or rutile are often enclosed in rock crystal (Pl. 97). Carbonaceous pigment may be embedded in calcite or andalusite (Pl. 86–7); quartz containing bituminous substance is called 'stink quartz' because of its smell when shattered.

4 PHYSICAL PROPERTIES OF MINERALS

ATOMIC STRUCTURE AND PHYSICAL PROPERTIES. The manifold electrical, magnetic, thermal, optical and other physical properties of crystals are closely linked with their molecular structure. Even crystal growth itself is subject to directional forces; for if the rate of growth were equal in all directions, then only spheres would develop from the centres of crystallization. Conversely, if minerals which have been cut into spherical shape are placed in solutions of compositions identical with their own, thus allowing further unrestricted growth, the crystal spheres will develop into crystals bounded by plane faces.

TRANSPARENCY. According to its transparency to visible light a mineral is called water-clear, transparent, translucent or opaque. Between these points there are innumerable intermediate stages. Finally, minerals may be translucent at their edges, i.e. only thin splinters are translucent, while larger individuals are translucent only at sharp crystal edges (e.g. zincblende and spinels).

DOUBLE REFRACTION OF LIGHT. In amorphous bodies, in gases, liquids and also in materials crystallizing in the cubic system, the light passing through is refracted only simply and also equally in

all directions. All other crystals split the incident light into two different components; they are therefore called *doubly refractive*. This double refraction is particularly striking with cleavage rhombs of clear calcite from Iceland, which is in consequence known as

◁ 26 HAUYNE blue cubic crystal with many faces and magmatically corroded edges, on scoriaceous lava. From Niedermendig, Eifel. Hauyne after the French mineralogist Hauy. $(Na, Ca)_{8-4}$ $[(SO_4)_{2-1}/(AlSiO_4)_6]$ cubic. Scale 6·6:1

27 HEULANDITE (stilbite) tabular crystals in a vug in augite-por- ▷ phyrite. The blood-red colour is caused by innumerable tiny specks of embedded goethite. From Palle-Alper, Fassa valley, S. Tirol, Italy. Heulandite after the British geologist Heuland. Stilbite after Greek *stilbe,* lustre. $Ca[Al_2Si_7O_{18}] \cdot 6H_2O$ monoclinic. Goethite after the poet and mineral collector Goethe. FeOOH orthorhombic. Scale 1·6:1

28 NATROLITE semicircular aggregates with radial acicular struc- ▷ ture and concentrically layered white and brownish coloration in a joint of greenish-grey phonolite. From Hohentwiel, near Singen, Hegau. Natrolite from Arabian *an natrun* (from Ancient Egyptian) and Greek *lithos,* rock. $Na_2[Al_2Si_3O_{10}] \cdot 2H_2O$ orthorhombic. Phonolite from Greek *phone,* sound and *lithos,* rock ('clinkstone'). An eruptive rock. Scale 2:1

29 CHALCEDONY blue fractured nodule with a yellow-brown crust; ▷ weathered from andesite. From Tresztya in Transylvania. After Chalcedon, town in Asia Minor. SiO_2 trigonal (cryptocrystalline). Scale 1·1:1

Iceland spar (Pl. 17). With many other doubly refractive minerals the phenomenon can only be seen with optical instruments. Minerals can be recognised and classified according to their refraction of light, their double refraction, the position of the optic axes in the crystal structure, their pleochroism and other optical properties. The optical properties are often correlated with chemical composition. Transparent minerals can be studied in thin-section, while opaque ore specimens are cut, polished on one surface and examined in reflected light. Such methods of investigation are among the most important in mineralogical work.

DISPERSION AND DICHROISM. The light passing through a crystal is refracted to different extents according to its wavelength (colour) — a phenomenon known as *dispersion*. The incident composite white light is analysed into its coloured components so that certain precious stones shine with the colours of the rainbow — they are 'on fire'.

Many crystals appear differently coloured in different directions. This is known as *dichroism*. If several colours can be seen in this way, one speaks of *pleochroism*. With transparent crystals of cordierite or tourmaline this property can even be seen with the naked eye. Pleochroism is a consequence of double refraction and so only occurs in non-cubic minerals.

LUSTRE. Minerals reflect light incident on them to very different extents. Lustre is a property depending on refraction and reflec-

30 CHALCEDONY, blue, stalactiform with yellow ends and rough ▷ pitted surface, incumbent on encrusting radial acicular limonite. From Naila, near Bayreuth, Bavaria. See Pl. 29. Scale 1·8:1

tion, and on a mineral's transparency and the state of its surface. Strong refraction almost invariably goes with strong lustre. The lustre of minerals is named by comparison with that of generally known materials, and so one speaks of adamantine, vitreous, resinous, greasy, waxy, pearly, and silky lustre — to name but a few examples. Minerals with metallic lustre are always opaque and often strongly reflecting. The degrees of intensity of lustre are described as splendent, shining, glistening, glimmering and dull or matt. The lustre of a mineral can be greatly impaired by efflorescence, thin coating films or incipient weathering at the surface. Earthy mineral aggregates always have a dull lustre.

THE COLOUR OF MINERALS. Apart from the rich variety of crystal forms, variety of colour is the most striking characteristic of minerals. In many cases the colour of a mineral is the natural colour of the substance of which it is made. Such *idiochromatic* minerals always have the same characteristic colour, from which they can readily be identified — yellow sulphur (Pl. 5), red cinnabar, green malachite (Pls. 9–10, 73), blue azurite (Pl. 75), lead-grey galena (Pl. 43). The colours are so characteristic for these substances as to be named after them: sulphur-yellow, cinnabar-red, malachite-green. But a much larger number of minerals owe their colouration to included foreign substances which are often present only in very small quantities. Alien ions present in the crystal lattice can cause colouration, and so can specks, scales or tiny grains of other minerals. Colours can also be produced by radioactivity (Pls. 1, 18, 27, 35, 50, 93, 97). Any mineral which owes its colour to foreign particles, i.e. any *allochromatic* mineral, may therefore appear in many different colours. With such minerals colour usually gives no clue to the mineral species. Differences therefore occur in the colour or in the strength of the colour of crystals of one and the same allochromatic mineral species which form part of a single crystal assemblage. And individual crystals which are differently coloured at each end are also known (Pls. 14, 19). Some mineral species occur in an amazing number of hues, e.g. fluorspar which forms not only transparent but also white, grey, wine-yellow, honey-yellow, brownish-yellow, rose, green, blue, violet and even darker (almost black) crystals. Quartz and chalcedony also form many coloured varieties. The colour of allochromatic minerals can in many cases be changed by heating, by bombardment with X-rays or other radioactivity or by exposure to ultra-violet light. Many coloured minerals, such as certain varieties of topaz or fluorspar, are even bleached by sunlight.

SPECIAL LIGHT EFFECTS. Light is reflected and diffracted by regularly intercalated foreign substances, by fine fractures or by twinning, and in this way distinctive light effects arise which are characteristic of many minerals. *Labradorescence* is a magnificent play of colours — predominantly in blue — which is particularly strong in labradorite (Pls. 2, 18). The colours change with the angle of incidence of the light, and they are due to fine twin lamination — i.e. to a multiple repetition of twinned intergrowths of very thin tabular crystal sheets — and to very small regular intercalations of crystals of haematite, magnetite or ilmenite. With opal the play of colours is called *opalescence* and depends on the special structure of this mineral, through which the light is reflected. As an opal is turned, bright rainbow colours become visible, and they migrate over the whole surface or shine out only from individual points (Pl. 20). Special colour effects can arise from thin surface skins, e.g. from oxidation products. These are the bright colours of thin sheets, and their presence in many minerals makes determination of the species difficult. Such phenomena can often be observed in copper indigo or covellite (Pl. 22).

STREAK. The colour of the powdered mineral on a white underlay is called its streak. It is often different from the colour of larger pieces and is often diagnostic of a given mineral species. Thus pyrite crystals are yellow, but their powder is greenish-black. Black zincblende has a leather-brown streak, a black limonite nodule a brown one, and blue-black haematite a blood-red one. Many classification tables are in consequence based on the streak colour, which can readily be ascertained by rubbing the mineral on the surface of an unglazed porcelain plate, a 'streak-plate'.

HARDNESS. Hardness also belongs among the physical properties which have long been in use as criteria for recognising minerals. In the ten stages of Mohs' scale of hardness, talc has hardness 1, gypsum 2, calcite 3, fluorspar 4, apatite 5, potash felspar 6, quartz 7, topaz 8, corundum 9 and diamond (the hardest mineral) 10. A mineral will scratch all those below it in the scale, and will itself be scratched by all that are above it. Minerals of hardness 1 and 2

◁ 31 CHALCEDONY bluish nodule with brown crust; layered red and white inside. Polished section. From Uruguay. See Pl. 29. Scale 1·5:1

32 AGATE layered, with circular brown patterns. Clear quartz on ▷ the top left. From Brazil. Agate after the river Achates in Sicily. A variety of chalcedony or cryptocrystalline quartz. Scale 0·9:1

33 DENDRITIC AGATE chalcedony mass with red- brown, ill-defined ▷ spots and black fern-like inclusions (dendrites): polished section. From Radschkot, Cathinar peninsula, India. A variety of chalcedony or cryptocrystalline quartz. Colouring due to inclusions of the hydroxides of iron and manganese. Scale 1·9:1

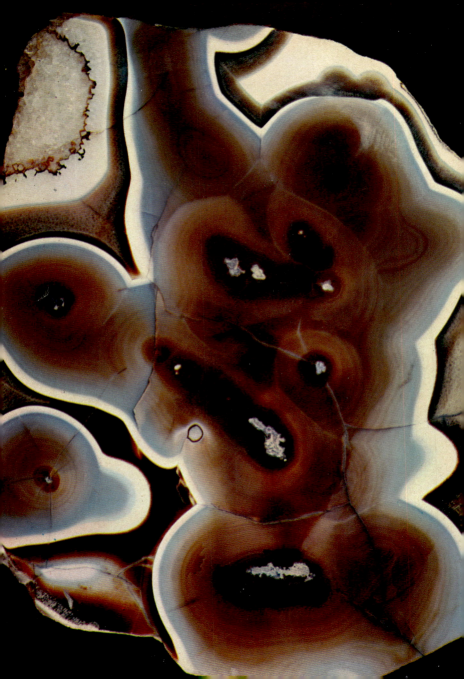

are soft enough to be scratched with the fingernail, and those of hardness 1 are, in addition, soapy to the touch, like talc or graphite. A knife will scratch all minerals up to hardness 5, while those of hardness 7 or over will scratch window-glass. The interval between each stage in this scale is not the same and is particularly large between corundum and diamond.

The harder a mineral, the more it retains its lustre when ground, and the less it suffers from the tiny sharp-edged quartz particles in the dust of the atmosphere. For this reason degrees of hardness above 7, which cannot be scratched by quartz, are called *gem-stone hardness*. Many minerals show definite differences of hardness on their faces according to the direction in which they are scratched. This is particularly the case with kyanite, which can only be scratched in one direction with the point of a knife. Mineral aggregates often appear to have a lower hardness than that of their individual component minerals, and the same is true of rocks. Many poorly cemented sandstones can be scratched with a knife, although they consist of grains of hard quartz.

DENSITY. Many minerals can be identified from their density merely by holding them in the hand. The minerals with the heaviest specific gravity are native gold and the platinum metals, which are about twenty times heavier than water. Most minerals that look rock-like, especially the silicates, have a density that lies bet-

34 MUSCOVITE spindle-shaped crystals with some transverse fractures developed in coarse potash felspar. From Lake Ilmen, Urals. So named because transparent mica sheets were imported from Russia. KAl_2 $[(OH, F)_2/AlSi_3O_{10}]$ monoclinic. Scale 1·4:1

ween 2·5 and 3·5. Ore minerals of metallic appearance lie considerably above these figures, and many sulphides and oxides of heavy metals have specific gravities of between 4 and 8. Most of the ores and minerals in veins, e.g. calcite, brownspar (an iron-bearing variety of dolomite), dolomite or fluorspar, range near 3, while barytes is appreciably higher (about 4·5).

CLEAVAGE is a crystal property depending on molecular structure and cohesive forces. Cleavage may be perfect, good or merely incipient. Normally a crystal can be cleaved more readily in certain crystallographic directions than in others. Rock salt, for instance, forms cubic cleavage planes, fluorspar octahedral and calcite rhombohedral ones. The more perfectly cleavage is developed, the planer and smoother are the cleavage surfaces, and the thinner the sheets that can be split off — among the thinnest are cleavage flakes of mica (Pl. 21) and gypsum. Minerals with such perfect cleavage readily form 'books' and their cleavage planes have a pearly lustre. Miners have named a whole series of minerals after their well-developed cleavage: calcareous spar, fluorspar, heavy spar, brownspar, iron spar or felspar. In contrast to crystalline minerals amorphous ones are without cleavage.

FRACTURE. When a mineral is shattered or broken open, apart from the plane cleavage surfaces, which follow fixed crystallographic directions, *fracture surfaces* may also be formed — although they are not likely to arise with minerals whose cleavage is good or perfect. According to the appearance of the fracture surfaces one speaks of conchoidal, smooth splintery, hackly, fibrous, even or uneven fracture.

CREATION AND DESTRUCTION OF MINERALS. Sparkling gem-stones and colourful crystal assemblages which were formed many millions of years ago in their mother rocks are for us a symbol of immortality — and they are indeed imperishable when measured by the short span of human life. But although minerals and rocks may appear indestructible to us, they form part of the cycle of change which includes all natural phenomena. New minerals are continually being formed at countless places in the earth, while others are destroyed, dissolved or reconstituted. Minerals can be formed in a great number of ways, from magmatic melts, from volcanic gases and vapours, from aqueous solutions of the most diverse origin, or from the recrystallization of solid matter. *Igneous rocks* with their richness of minerals of different kinds are formed from the cooling of magmatic melts at depth or of lava brought up to the surface by active volcanoes. When they are exposed on the surface they are subject to weathering, erosion and solution. Water transport removes materials taken into solution and eroded particles — boulders, pebbles, gravel, sand and mud. The breakers of the sea or the pounding of lake waves can also erode, and comminuted material can be further transported by wind, which can abrade as well as carry. Enormous amounts can also be transported by glacier ice in high mountains, in the Arctic and Antarctic. Material that has been mechanically broken down and transported by streams, currents, ice or wind is deposited, and thus new rocks are formed. When evaporation occurs, material in solution can be precipitated as mineral salts. And finally organisms also contribute to the formation of new rocks and minerals. Rocks formed in various ways from deposition on the earth's surface, on the sea floor, in estuaries, rivers or lakes, are *sediments*.

Sedimentary masses are normally in the first instance poorly consolidated, and on compaction quite new mineral species may be formed, or existing ones recrystallized. Like all the rocks of the earth's surface, even sediments that have just been formed may be subject to immediate erosion. As a result of mountain-building processes or the uprise of magmatic masses, igneous rocks or sediments may be exposed to increased pressure or temperature, and then their constituent minerals are reconstituted, and *metamorphic rocks* and minerals are produced.

Finally, if a whole sequence of rocks is lowered to great depth or if it interacts for long enough with magmatic melts, it may itself be melted. The first minerals to liquefy will be those with lowest melting point, and the process can continue until the whole sequence has gone over to a magmatic state — *ultrametamorphism.*

◁ 35 EMERALD (beryl coloured green by small amounts of chromium) long prismatic, green, transparent crystals in mica-schist. From Takovaya valley, near Swerdlovsk, Urals. Greek *smaragdos.* $Be_3Al_2 [Si_6O_{18}]$ hexagonal. Scale 1·4:1

36 TOURMALINE section cut vertical to the principal crystallographic axis (the long axis) through a pencilled multi-coloured crystal. The colour distribution corresponds to the trigonal symmetry. From Madagascar. See Pl. 19. Scale 0·7:1 ▷

37 SCHEELITE yellow-brown warty crystals resting on a larger quartz crystal which is partly covered by fine scales of red haematite. From Schlaggenwald, Bohemian Erzgebirge. After the Swedish chemist Scheele. $Ca[WO_4]$ tetragonal. Quartz SiO_2 trigonal. Haematite Fe_2O_3 trigonal. Scale 2·5:1 ▷

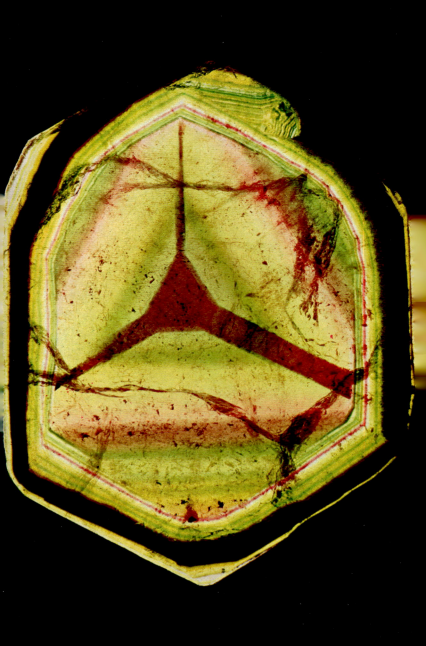

The cycle then begins anew. During liquefaction there may be interaction between the uprising magma and the country rock. Rocks only partly melted can have their liquid fraction squeezed out, and it is then free to crystallize elsewhere. Bodies of rock from the envelope overlying a magma may be engulfed in it and melted. Solutions emanating from magmatic melts can migrate far into the country rock and deposit their mineral contents in remote places.

Minerals can thus be formed in a great variety of places under widely varying conditions of temperature, pressure and in many different rock environments. New minerals are continually being formed and others destroyed in magmatic environments, on the deep-sea floors, in the roots of continents and at the exposed surface. Admittedly, it only rarely happens that conditions are right for the formation of large crystal groups which can grow freely. There is not always enough space for large crystals to grow without distortion, and in many places where a large number of crystals are growing simultaneously, the supply of material is insufficient to sustain them.

MINERAL PARAGENESIS. It only rarely happens that a whole rock mass is formed of a single mineral species. Normally several kinds of minerals go to make a rock or an ore bed. But over the whole earth certain minerals are again and again found in association — in composite wholes that vary according to the conditions of ori-

38 MOLYBDENITE a hexagonal single crystal of flat tabular habit ▷ that has grown on granite. From Wakefield, Quebec. Molybdenite from Greek *molybdos,* lead, because the soft mineral stains when touched, just as lead does. MoS_2 hexagonal. Scale 5·8:1

gin. Minerals that have originated together as a result of the same formative process are called *paragenetic,* and the resultant mineral assemblage is a *paragenesis.* Many parageneses are of world-wide distribution and often occur in quantities of many hundred cubic miles. Thus granite, the commonest plutonic rock, is a combination of potash felspar, plagioclase, quartz and biotite. Other parageneses are to be found at only a few places, e.g. emerald in mica-schist (Pl. 35). Finally there are conditions of formation which — as far as our observations go — are rare or even unique, and which have produced correspondingly unusual parageneses. Many minerals are met with in numerous parageneses, such as the ubiquitous quartz, calcite or pyrite (fool's gold). Mineral species which — like pyrite — occur in magmatic, sedimentary and metamorphic parageneses are of course not diagnostic of a certain mode of origin. Other minerals are formed only under quite specific temperature and pressure conditions, or only in the presence of certain country rocks or when certain other substances are also present in the solution. Where such minerals are found these specialised conditions must have obtained, and at such localities minerals that normally occur only very rarely may be present in large quantities. Mineral species that are formed together in one or another of the various modes of origin are not only built of certain chemical elements, but often also contain specific tracer elements. Thus the minerals which form any given assemblage never represent a random accumulation, and the number of combinations which occur naturally is not very large. Minerals from which metals or other elements can be extracted are grouped as *ore-minerals* or ores. A particular mineral species is often found only in small amounts at a given locality. When greater mineral concentrations rich in ores or other useful minerals occur in the crust, one speaks of ore deposits, salt deposits, etc. In such cases the mineral-forming processes were particularly powerful or of unusually long duration.

SYNGENETIC AND EPIGENETIC MINERAL FORMATION. Minerals formed by the same processes as the rocks in which they are embedded are called *syngenetic*. Such deposits are of the same age as the rocks adjacent to them. Most sedimentary mineral concentrations belong to this group, and such deposits — and salt deposits too — are therefore sometimes called synsedimentary. In contrast with these are the minerals which may have originated very long periods of geological time after the rocks in which they occur. Such mineral occurrences, e.g. ore veins in rocks of completely different and unrelated type, are called *epigenetic*. Petrology and ore research, important branches of mineralogy, are much concerned with the history of rocks and ore deposits, the materials of which they are made and the processes by which they are reconstituted.

WHERE MINERALS ARE FORMED

In mineralogical textbooks minerals are mostly classed according to their chemical composition and their molecular lattice structure. In the coloured plates of this book minerals are grouped according to their place in paragenetic groups, i.e. according to their natural occurrence in the major realms of mineral formation: magmatic, sedimentary and metamorphic. In this grouping mineral species whose chemical composition and lattice structure are closely allied are illustrated in different sections of the book, and many minerals have been given several illustrations. The purpose of all this is to stress that no mineral has arisen in isolation and that the paragenetic occurrence of certain minerals obeys laws of its own. Let us therefore scrutinise the major realms of all processes of geological deposition and reconstitution with the purpose of ascertaining which precise conditions will favour the growth of particularly fine minerals and well-developed crystals.

◁ 39 GALENA angular fragments of country rock (gneiss), with thin crusts of galena, lying in a mass of milky vein quartz. Breccia ore. From Anton mine, Wieden, Black Forest. Italian *breccia*, rock cliff. Galena PbS cubic. Scale 1·4:1

1 MAGMATICALLY FORMED MINERALS

The term *magma* designates the hot liquid masses of the deeper crust from which magmatic rocks are formed. When a magma solidifies at depth, a plutonic rock is formed; but if the liquid masses penetrate to the earth's surface they are called *lavas,* which on solidification form what are perhaps the most important of the volcanic rocks — although this category also includes the pyroclastic rocks (ashes, tuffs, etc. noted below).

Mineral crystallization from volcanic lavas occurs within the temperature range of 1100°C to 700°C. Although the mineral parageneses of magmatic rocks, and of the highly metamorphic

◁ 40 Native COPPER dendritic aggregate of distorted tetrahexahedra. From Lake Superior, Michigan. Latin *cuprum,* the older form being *aes cyprium,* Cypriot ore, from its occurrence in Cyprus. Cu cubic. Scale 2·1:1

◁ 41 Native GOLD leaf-shaped crystals, some bent, grouped in rosettes, grown on quartz. From Rosia Montana (Verespatak) in the Transylvanian Erzgebirge. Gold Au cubic. Quartz SiO_2 trigonal. Scale 4:1

42 Native ARSENIC nodular and layered, covered with a film of ▷ yellow realgar. White coarse calcite between the nodules. From Pribram, Bohemia. Greek *arsen,* strong, because it was used as a tonic. As trigonal. Realgar from Arabic *Rah-ja gar,* powder of the cave. AsS monoclinic. Calcite $CaCO_3$ trigonal. Scale 1·7:1

rocks akin to them in composition, comprise only relatively few mineral species, these are the species of which the earth's crust is very largely made up. In fact to a depth of 10 miles they contribute 95 % of it.

The solidification of magmas at depth begins with early *orthomagmatic crystallization,* that is the precipitation of high melting-point minerals of titanium, phosphorus, chromium, vanadium, platinum and kindred metals. Among the most important of these minerals are magnetite, titanomagnetite, ilmenite, rutile and other spinels, as well as native metallic platinum. The enormous accumulations of titaniferous iron ores are among the greatest metal concentrations in the crust. Nickel pyrites, chalcopyrite and other sulphides may become separated, migrate from the melt as droplets and likewise form considerable accumulations. Iron can migrate from iron-rich magmas containing phosphorus, and can crystallize out, away from the rest of the magma, forming large ore deposits of magnetite (magnetic iron ore) and apatite. The mineral parageneses of early orthomagmatic crystallization are restricted to certain plutonic rocks; these are rich in iron and magnesium, relatively poor in silica, and often free of felspars. Within this sphere different mineral combinations can arise if the liquid is squeezed away from the crystal sludge, so that both these magma fractions crystallize in separate places. The type of crystals that grow freely in cavities (called vugs) cannot be expected among the products of this stage of crystallization, for no open cavities can form in magmatic bodies, which are under great pressure from all directions. There are, however, often large crystals in these plutonic rocks which have solidified very slowly.

Diamonds are found in silica-poor host rocks brought up from vents or rifts from very great depth. Diamond can only be formed under very high pressure and so cannot originate in ordinary plutonic rocks.

84

As the temperature of the magma falls, early crystallization is succeeded by the main orthomagmatic stage — the stage at which the mineral components of normal igneous rocks are formed. Generally speaking, the more slowly the crystallization of a magma proceeds, the larger are the minerals formed. Deep-seated (plutonic) rocks are usually coarsely crystalline because of slow crystallization. In massive plutonic rocks formed under high pressure the superincumbent rocks prevent any rapid escape of the gases dissolved in the magma which are liberated as it crystallizes. Conversely in volcanic rocks, which cooled at the earth's surface or at shallow depths, the rock minerals are normally fine-grained and often so small that they cannot be distinguished with the naked eye. These fine-grained dense rock masses, such as the basalt sheets spread over enormous areas, are devoid of minerals of great size, unless they contain vesicles or amygdales.

Many igneous rocks contain sizable and also quite well-formed minerals which lie embedded as phenocrysts in a fine-grained or dense groundmass. These phenocrysts formed slowly at great depth, whereas the groundmass solidified quickly when, at a later stage, the magma rose up and erupted at the surface. Those minerals which formed down in the magma reservoir are called intratelluric phenocrysts (Pl. 24). Volcanic bombs, lapilli and fine-grained volcanic ashes derive from gas-rich lavas which are dispersed by their high internal pressure on reaching the earth's surface. Such deposits are known as volcanic tuffs.

Dyke rocks occupy an intermediate position between plutonic rocks and volcanic lavas and form a variety of rock types. They derive from magma which has penetrated into clefts and joints. The majority of dyke rocks are of basaltic composition and the dykes in question have in many cases acted as feeding channels for surface lava of the same composition. It will be appreciated that dyke rocks can show the same range of composition as lavas. There

are however, two main kinds of rock type found only in dykes (often closely associated with granitic plutons): these are the light-coloured aplites, consisting mainly of quartz and felspar, and the dark lamprophyres. Dyke rocks generally have chilled and some-times glassy margins where they abut against the country rocks. If extruded lavas cool very quickly, considerable amounts of *volcanic glass* may be formed. Acid lavas in particular often form sometimes quite extensive masses of scoriaceous, porous or splintery pumice stone or amorphous glass. Rock glass is absent from pluto-nites which have crystallized much more slowly, often throughout

◁ 43 GALENA crystals with many faces — combination of the octa-hedron, cube and rhombododecahedron — with brass-yellow chalcopyrite and brown translucent fractured siderite rhombs. Encrusting bournonite at the top. From Pfaffenberg mine, near Neudorf, Harzgerode, Harz mountains. Latin *galena,* lead ore. PbS cubic. Chalcopyrite $CuFeS_2$ tetragonal. Siderite from Greek *sideros,* iron. $FeCO_3$ trigonal. Bournonite after the French crys-tallographer Count Bournon. $PbCuSbS_3$ orthorhombic. Scale 1·7:1

44 ZINCBLENDE (sphalerite) honey-brown and darker crystals with ▷ many faces, dodecahedral habit and adamantine lustre, on ag-gregates of transparent quartz crystals. From Schemnitz, Slovak Erzgebirge. Sphalerite from Greek *sphaleros,* deceptive. ZnS cubic. Scale 3·6:1

45 GOLD aggregate of native gold on quartz crystals. From Rat- ▷ hausberg, near Bad Gastein, Salzburg, Austria. Gold Au cubic. Scale 4·1:1

whole geological epochs. Glassy masses like obsidian and pitch-stone are therefore not minerals but supercooled rock melts of variegated chemical composition which have solidified as glass. In rock glass formed from acid lavas which cooled quickly are found small skeletal crystals. These microliths — fine rodlets or spherulites — were unable to develop into larger crystals because the lava solidified so quickly. In the course of geological time, rock glass is gradually transformed. As it ages it becomes devitrified.

Among the many known types of igneous rocks — they derive from all geological periods — two are greatly predominant: granites and basalts. Basalt is the most widely distributed volcanic rock, and is formed from melts which rose quickly to the earth's surface, often from very great depths. On the other hand granites and kindred rocks are the commonest of the plutonites. They are found in the cores of folded mountain chains and form enormous rock masses in all continents. The variety of the remaining igneous rocks depends not only on differences in the composition of the magma, and on the manifold geological relationships during cooling, but also on the removal or migration of crystal material during cooling — *crystal differentiation* as it is called. Thus early precipitated crystals of high specific gravity may sink through the magma, while lighter crystals rise, resulting in a layered sequence. On the other hand many early precipitated minerals may be changed or melted by the magma at a later stage of crystallization, while the chemical composition of the magma is continually changing as crystals are formed.

46 Native SILVER tree-like assemblage of many small cubic crys- ▷ tals, partly covered by white barytes. From Anton mine near Schiltach, Black Forest. Silver Ag cubic. Barytes $BaSO_4$ ortho-rhombic. Scale 3·6:1

Because of its high melting point olivine, named after its olive green colour, is among the earliest minerals precipitated in an igneous rock. Sometimes olivine nodules, surrounded by a scoriaceous crust, are ejected as bombs during volcanic explosions (Pl. 23). These nodules consist of debris of many intergrown olivine granules of compressed habit, and of individual little crystals of emerald-green chrome diopside.

The rock-forming minerals of the igneous rocks include dark components rich in iron and magnesium, and light-coloured ones which contain mainly silica and aluminium. The principal dark minerals of igneous rocks are olivine, the augites and hornblendes and dark mica (biotite). The light-coloured components include quartz and the felspars. Magmatic rocks are classified according to their essential components, and in this connection the felspar content is of particular significance. Minerals present in addition to the essential components and therefore represented in smaller quantities are called accessories. Magmatic rocks with a high silica content, e.g. granites, are called acid. Acid igneous rocks usually contain free quartz, often in considerable quantities. Rocks with an intermediate silica content are termed intermediate, and quartz-free rocks, with a low silica content, are known as basic. The most widely distributed basic igneous rocks are the basalts.

In certain types of silica-poor igneous rocks the felspars are replaced by a special group of minerals, the *felspathoids*. The most important members of this mineral group are leucite, nepheline, sodalite, nosean and hauyne. Since felspathoids plus quartz would form felspars, they can only occur in quartz-free rocks. In some rock types leucite forms crystals which may be as large as cherries and which are admirably developed with many faces (twenty four is a common number). Like the other felspathoids leucite occurs only in its position of growth. In some rocks hauyne crystals show indistinct corroded margins; the crystals, which were already precipitated in the magma reservoir, were at a later stage attacked by the hot melt and in part redissolved.

Whereas deep-seated rocks crystallize to form irregular grains and have a massive unorientated texture, lavas which flow as streams and as sheets over the earth's surface often show a *flow texture,* with the minerals in some special alignment. Like logs in a river, elongated prismatic and lath-shaped crystals (e.g. felspars) are orientated in the direction of flow. In deep-seated rocks the minerals are precipitated in a definite sequence, depending on the composition of the magma. In granite rocks the sparsely represented accessory minerals have been regarded as the first to form — zircon, magnetite, ilmenite, haematite, rutile and apatite (important as a source of phosphorus). Then follow hornblende and biotite, then the felspars and finally quartz, which has to accomodate itself in the interstices of the minerals formed before it. Quartz only occurs in the igneous rocks that are fairly rich in silica. In the remaining deep-seated and eruptive rocks all the silica is contained in the silicate minerals. The earlier a mineral crystallizes in a magmatic rock, the more readily can it assume the crystal form proper to it.

The commonest minerals — the ones which to a large extent determine the character of most igneous rocks — are the *plagioclases.* They form a mixed-crystal series with lime and alumina content increasing as the soda-silica content drops: albite, oligoclase, andesine, labradorite (Pls. 2, 18), bytownite, anorthite. Which of these plagioclases will be precipitated in a given case will depend on the chemical composition of the magmatic melt. The exact nomenclature and classification of the igneous rocks is to a large extent based on their felspar content — both on the type of felspar minerals and on the proportion which they form of the whole. While the potash felspar of deep-seated rocks is mostly developed as orthoclase (Pl. 19), in quickly cooled surface lavas it appears as glassy-

looking sanidine. Biotite, the commonest of the micas, usually dark brown or black, occurs in thin 'leaves' or scaly masses. Like all micas biotite — ferromagnesian mica — is monoclinic, but often forms pseudohexagonal crystals six-sided in outline (Pl. 25). Augite (Pl. 24) and actinolite (Pl. 94) also belong among the dark rock-forming minerals. At temperatures above 573°C quartz crystallizes in a high temperature form which can be recognised from the hexagonal bipyramids of squat habit. Such quartz crystals are characteristic phenocrysts in acid lavas such as rhyolite and quartz porphyry. The phenocrysts are often corroded and *embayed* by the magma. Below 573° quartz in deep-seated rocks forms prismatic

◁ 47 PROUSTITE (light red silver ore) bright cinnabar-red translucent crystals of pyramidal habit and strong lustre. From Chanarcillo near Copiapo, Chile. Proustite after the French chemist Proust. Ag_3AsS_3 trigonal. Scale 4·6:1

48 ZINCBLENDE stalactiform banded aggregates. From Kobelsberg ▷ near Wiesloch, Württemberg. Zincblende ZnS cubic. Scale 1·7:1

49 CHALCEDONY as a yellow crust covering quartz crystals, and in ▷ turn covered by a further brown crust of chalcedony stained by limonite. From Haizamkhab, India. SiO_2 trigonal. Scale 3·1:1

50 CHRYSOPRASE (green chalcedony) bright leek-green specimen ▷ with uneven colouration, showing fracture surfaces. The colour derives from admixture with a little nickel. From Kosemütz, Silesia. Greek *chrysos*, gold, and *pras*, leek. Cryptocrystalline quartz. SiO_2 trigonal. Scale 1·8:1

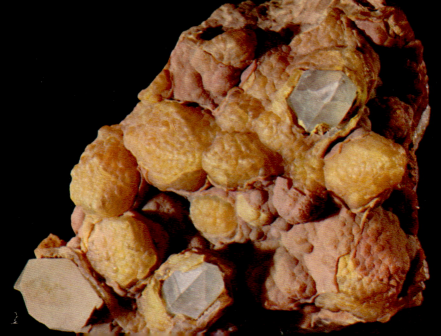

columnar crystals, often extended to form long pencils (Pl. 102) with pyramidal ends. Minerals like this which develop special modifications or crystal forms according to the temperature are called *geological thermometers*. Some minerals allow us to infer the pressure conditions obtaining at the time of their formation, and these are accordingly named *geological manometers*. Well-formed minerals from the orthomagmatic stage of crystallization are rarely found in deep-seated rocks. Normally the components crystallized out simultaneously and interfered with each other's growth. On the other hand the phenocrysts in many lavas are quite large and often show perfect crystal form. So long as these rocks are fresh it is not easy to remove the various phenocrysts of augite, hornblende, felspar or felspathoid. But once the rock has been slightly weathered, the idiomorphic phenocrysts can often easily be taken from the friable groundmass without damaging them. Volcanic tuffs are also often good sources of well crystallized magmatic minerals and sometimes contain fine specimens of hornblende, augite or biotite. Well crystallized intratelluric minerals are present in many volcanic screes in such large numbers that these screes are called *crystal tuffs*.

51 BLENDE brown banded intergrowth of zincblende (sphalerite) ▷ and wurtzite. Also some rather dark galena. A finely fibrous growth forming a layered and reniform encrustation. The surface has been cut and polished. From Stolberg, near Aachen. Zincblende ZnS cubic. Wurtzite after the Alsatian chemist Wurtz. ZnS hexagonal. Galena PbS cubic. Scale 1·2:1

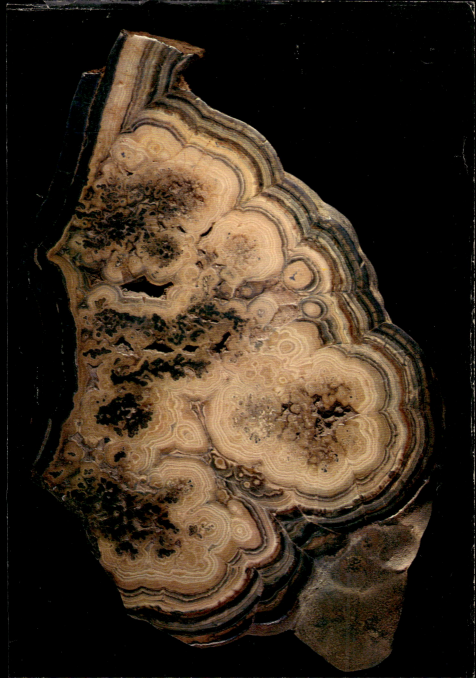

Active volcanoes bring up enormous quantities of steam and other gaseous substances and vapours. Volcanic exhalations also include volatile compounds such as carbon dioxide, sulphur dioxide, and compounds of chlorine, fluorine and boron, most of which pass into the atmosphere. Because the pressure at the earth's surface is so low gaseous products also escape from the lavas themselves, so that vesicles are formed in the melt which are preserved if it has already become viscous. The explosively liberated volcanic gases react with each other, with the melt and also with the minerals that have already crystallized, and in this way numerous new minerals, often with well-developed crystals, may be formed. Points at which hot vapours are exuded from active volcanoes are called *fumaroles*. Exhalations at lower temperatures from quiescent or extinct volcanoes are termed *solfataras*. Finally *mofettas* − giving off carbon dioxide − or *thermal springs* can continue for some time after the cessation of volcanic activity. Apart from sulphur and sassoline (boric acid), there are several dozens of mineral species formed from fumaroles and solfataras, according to the temperature and chemical composition of the fumes exuded. Particularly interesting parageneses which include rare minerals arise where volcanic vapours are discharged into the sea or react chemically with carbonate country rocks. All these minerals are normally perfectly formed although only of moderate size.

Vesicles in basalts, phonolites, melaphyres, diabases and other volcanic rocks frequently contain zeolites. The numerous minerals of this large family occur predominantly as white or colourless crystals or as brownish coloured aggregates. According to the crystal habit one can distinguish granular, scaly and fibrous zeolites. A widely distributed mineral of the zeolite group is natrolite, a fibrous zeolite found in various forms and aggregations in joints and vugs (Pl. 28). A small amount of limonite gives this mineral a yellow or ochre-brown colour. Heulandite (Pl. 27), a scaly zeolite, is coloured red by fine scales of goethite (hyroxide of iron). Mineral-bearing cavities in vesicular lavas are called amygdales. Many cavities are completely filled with chalcedony, in others the layers of chalcedony are confined to the walls, while quartz or amethyst crystals project into the hollow interior. Calcite is a common associate in these chalcedony vugs, and goethite, micaceous goethite and haematite sometimes coat the crystals. *Lithophises* are gas-cavities of irregular shape. They occur in rhyolites, quartz-porphyry and other siliceous lavas and are often filled with chalcedony, agate, tridymite, quartz, haematite and carbonates.

Chalcedony is a cryptocrystalline modification of quartz formed from siliceous gel. *Cryptocrystalline* describes mineral aggregates consisting of such small individual crystals that they can only be distinguished under the microscope – in contrast to phanerocrystalline aggregates, composed of individual minerals that can be seen with the naked eye. Thus chalcedony is an aggregate of tiny quartz fibres: in some varieties these are embedded in an opal mass. The finely crystalline chalcedony minerals appear outwardly compact and homogeneous, and occur in an amazing number of colours and forms. Chalcedony, which often has a waxy lustre, forms coarse masses with botryoidal, reniform surfaces. It may be nodular, stalactitic (Pl. 30) or encrusting; translucent or black and opaque, and may occur in bright colours and in a number of complicated growth forms (Pls. 29, 31). Coloured varieties of chalcedony have long been specially named, since they are prized as gem-stones. Masses of chalcedony built up from different coloured layers are called *agates*. According to the way in which the chalcedony layers are grouped one speaks of

'landscape agates', 'circular agates', and so on (Pl. 32). An agate shattered by tectonic stresses and subsequently recemented with quartz material is a 'ruin agate' (more properly called brecciated agate). The forms assumed by agate banding are well-nigh inexhaustible. In polished thin plates or amygdales that have been sawn through, the translucent white chalcedony layers with the coloured bands form attractive patterns. Many melaphyres and similar lavas contain amygdales filled with amethyst (Pl. 1). Often the pyramid-shaped crystals rest on a basement of chalcedony of different colour. In different localities the colour of amethyst varies between light and dark violet, with a tinge of blue or red. The most prized amethysts are the dark violet-blue ones. Amygdaloidal masses may weigh several hundredweight, and normally only the terminal faces of the crystals in them are of

◁ 52 TETRAHEDRITE (grey copper) steely blue tetrahedral crystals with colourless translucent quartz. Some limonite top right. From Kapnik, Transylvania. Greek *tetraeder,* four-planed. $(Cu_2, Ag_2, Fe, Zn)_3Sb_2S_6$ cubic. Limonite $FeOOH$ orthorhombic. Scale 1·8:1

53 BARYTES (heavy spar) in cockscomb form of tabular crystals ▷
54 arranged nearly parallel to one another. Also pale violet cubes of fluorspar forming (top left) a crystal pavement. A distorted fluorspar crystal lies centrally. Small colourless quartz crystals. Between the barytes lamellae thin encrusting red brown limonite. From Tannenboden mine, near Wieden, Black Forest. Greek *barys,* heavy. $BaSO_4$ orthorhombic. Fluorite from Latin *fluor,* liquid, because used in metallurgy as a flux. CaF_2 cubic. Limonite $FeOOH$ orthorhombic. Scale 1·7:1

gem-stone quality, while prismatic faces nearer the walls are clouded and opaque. The crystal ends (pure in colour and ready for polishing) which have been carefully removed are called pointed amethysts. To remove agate amygdales from consolidated lava is both tedious and difficult to achieve without damaging them. The best localities for collecting specimens are where weathering has loosened the grains of the rock in which the amygdale is embedded, or where streams and rivers have deposited the weathered-out amygdales as placers.

From hot springs siliceous sinter (*geyserite*) is deposited. *Opal* is also a hardened gel mass (of dilute silicic acid). It occurs as an infilling of rounded cavities in basalts, andesites and trachytes, and as an encrusting deposit in the joints and clefts of other volcanic rocks of recent origin. Glass-opal or hyalite is crystal-clear and quite transparent. Gem opal (Pl. 20) is a coveted precious stone, with its bright play of colours. Stones with blue or green coruscation are specially prized. Fire-opal is a striking red. Country rock impregnated with opal-substance is called 'mother of opal' or opal matrix. Opals do not keep their beauty under all conditions. In a dry atmosphere they lose water by evaporation, and this impairs their fine colours. But loss of water and fading of the colours can be prevented if the stones are occasionally soaked in water or placed in a raw potato.

55 RHODOCHROSITE reddish layers of different colour round a ▷ brownish-black compact mass of psilomelane. From Restaurodora, Argentine. Greek *rhodochroos,* rose-coloured. $MnCO_3$ trigonal. Psilomelane from Greek *psilos,* bald, and *melas,* black. MnO_2 orthorhombic. Scale 1·4:1

A far greater number of mineral species are formed in the *post-magmatic phases of deposition* than in the igneous rocks. As igneous rocks crystallize at depth the residual magma becomes enriched in the volatile constituents which were dissolved in it from the first. The superheated aqueous silicate melts contain quite considerable amounts of highly volatile compounds whose elements cannot be accomodated at all (or only to a very small extent) in the rock-forming minerals of orthomagmatic crystallization. Examples are lithium, beryllium, zirconium, boron, niobium, tantalum, thorium, uranium and the rare earths. Hence the internal pressure of these superheated melts is very high, and the residual solutions of cooling magmas, enriched as they are in these volatiles, raise the pressure still higher.

Post-magmatic minerals are only found where superincumbent rock-masses have prevented the rapid escape of the superheated melt solutions. In the case of volcanic eruptions at the surface, the gases and vapours which pass into the atmosphere are at a pressure considerably lower than that which obtains in deep-seated rocks, and so lavas are unable to form pegmatitic or pneumatolytic minerals.

At temperatures between 700°C and 550°C mixtures of coarse-grained silicate minerals are formed – so-called *pegmatites*. This stage of crystallization and the minerals formed in it are accordingly both called pegmatitic. Accumulations of pegmatitic minerals form masses or layers with sharp boundaries, especially in the roof and marginal zones of magmatic rocks. The great majority of pegmatites are associated with the dominant plutonic rocks, i.e. granites and kindred types. Pegmatites are also characteristically associated with the more limited occurrences of syenitic rocks, but normally speaking one does not associate typical pegmatites with basic rocks. The commonest pegmatites are coarse-grained accumulations of quartz and felspar, often with muscovite (Pl. 21) and tourmaline (Pl. 19). Rarer minerals are often also present and a major fraction of many pegmatites consists entirely of quite rare mineral species. Pegmatitic minerals are not normally precipitated in a definite sequence, but crystallize almost simultaneously. A striking kind of intergrowth of quartz with orthoclase or microcline, reminiscent of written characters, is called graphic granite. Pegmatite masses are often zoned in structure, the minerals in the centre of the body being larger and of different species than those nearer the margins. In the marginal regions of granite massifs that have been permeated by pegmatitic solutions small cavities are formed which are described by the Italian mining term of *miaroles*. If the small cavities in such a miarolitic rock are very numerous, the rock acquires the appearance of sugar-grains. In the small cavities may be found rosettes of muscovite, tiny quartz crystals and haematite sheets. The larger pegmatitic cavities are well-known for their particularly fine crystals. The minerals emplaced in compact coarse-grained pegmatite masses and projecting freely into the centre of the cavities often consist of transparent or beautifully coloured well-terminated crystals which are much prized as polishable gems. Thus there are not only pegmatites which provide valuable raw materials for industry, but also gem-stone pegmatites. Emerald, which ranks high among gem-stones, can only form under very special conditions, i.e. when a beryllium-rich pegmatitic melt crystallizes in a rock environment containing chromium. Then, instead of turbid-white or grey-green beryl (the mineral of which emerald is the gem variety), a dark green transparent emerald may be formed.

Pegmatite minerals that have crystallized from hot aqueous residual melts have had conditions particularly favourable to their growth, for these melt solutions are not only very mobile, but also initiate few and well-spaced centres of crystallization, with the result that very fine and large crystals can grow from each centre. Often they penetrate far into the country rock. In pegmatites minerals are therefore usually very well-formed, and the extremely large crystals found in nature – including the very largest known – are pegmatitic. Examples include real giants of potash felspar (up to 180 tons), muscovite, phlogopite, spodumene and beryl (up to 100 tons). Apart from their exceptionally large crystals pegmatites are important for their very fine specimens of rare minerals, e. g. minerals formed from the elements cerium,

◁ 56 REALGAR (red) and orpiment (yellow). White calcite. From White Lake mine, near Manhattan, Nevada, U.S.A. Arabic *Rahja gar,* powder of the cave. AsS monoclinic. Orpiment from Latin *aurum,* gold, and *pigmentum,* colour. As_2S_3 monoclinic. Calcite $CaCO_3$ trigonal. Scale 1·5:1

57 MANGANESE SPAR (rhodochrosite) intergrown crystals of pine- ▷ cone shape and pointed, scalenohedral appearance, raspberry red and translucent, growing in a vug. From Herdorf near Siegen, Germany. $MnCO_3$ trigonal. Scale 4·7:1

58 CELESTINE bushy radial aggregates of colourless crystals with ▷ yellowish sulphur, both grown on a base of yellow-brown stalactiform calc sinter. From Roccalpalumba Mine, Palermo, Sicily. Latin *caelestis,* sky-blue, which is sometimes this mineral's colour. $SrSO_4$ orthorhombic. Scale 1·4:1

scandium, yttrium and other rare earths. Pegmatites with unique mineral combinations are known, and many mineral species are restricted exclusively to pegmatitic parageneses. Among the few pegmatite minerals which are not found as well-formed crystals is rose quartz, which forms compact crystalline masses broken by fractures. If the temperature of the residual magma continues to fall, then there is a further accumulation of volatile constituents, giving solutions rich in gases and in steam. In the range of 550°C to about 400°C lies the realm of those post-magmatic parageneses called *pneumatolytic*. Whereas quartz, potash felspar and muscovite (as well as many other minerals) are particularly characteristic of the granite pegmatites, silicates are not so important in pneumatolytic parageneses, although topaz, lithium mica and tourmaline (the commonest boron mineral) are typical silicate minerals of this stage. Mineral assemblages of pneumatolytic origin characteristically contain tinstone (cassiterite), wolfram, scheelite (Pl. 37) and molybdenite (Pl. 38).

The highly reactive pneumatolytic mineral solutions can penetrate into limestone and other carbonate country rocks even more powerfully than pegmatitic liquids, and 'boil off' there. In this manner many contact-pneumatolytic parageneses surprisingly rich in minerals are formed, and they include ore deposits which are world-famous for their variegated minerals and exceptionally well-formed crystals. In such parageneses silicates of lime, magnesium and iron are often found, and also valuable metal ores. Com-

59 BARYTES (heavy spar) tabular white crystals with a yellow-   ▷
    brown coating of limonite associated with water-clear cubes
    of fluorspar. From Clara mine, near Oberwolfach, Black Forest.
    See Pl. 53–4. Scale 1·6:1

pact masses of calc-silicate minerals which are closely inter-grown are called skarns, after an ancient Swedish mining term.

Pneumatolytic residual solutions often crystallize as veins, i.e. they crystallize as infillings of fissures in other rocks. These solutions can also attack their granitic parent rocks, or other similar rocks in the environment, and metamorphose or impregnate them. Transformations involving the formation of massive tourmaline or topaz are common. Rocks in which the minerals have been largely converted to tinstone used to be called 'greisen' by miners, while partial alteration, giving rocks consisting of about equal amounts of fine-grained granite and tin-ores, were known as 'peach'. Pneumatolytic minerals too are usually idiomorphic and often form crystals of quite exceptional beauty, although they do not attain the enormous size of pegmatite crystals. A number of minerals are known which occur only in pneumatolytic parageneses.

There are numerous transitions between the two post-magmatic stages of mineral formation, and the two are therefore often taken together, and one speaks of mineral assemblages of *pegmatitic-pneumatolytic* origin. An example of such an assemblage is the 'tin suite'. Miners in the Erzgebirge have long known that certain minerals are apt to occur with the tin ore they are seeking (cassiterite) – for example scheelite (Pl. 37), wolfram, zinnwaldite, tourmaline – and they called these minerals the 'tin suite'.

If the temperature of the residual magmatic liquid drops further, once the critical temperature is passed, a hot aqueous solution is formed, whereas previously the mineral solutions were in a super-critical state of low viscosity. The critical temperature for pure steam is 374°C (at a pressure of 225 atmospheres). The figure is somewhat lower for natural magmatic solutions and fluctuates according to the nature and amount of the material in solution. From these aqueous mineral solutions of magmatic origin the *hydrothermal mineral parageneses* are deposited. The solutions force their way up into joints, faults and clefts, and according to the space available form narrow stringers or thick mineral veins. In addition the mineral solutions may permeate and impregnate porous rocks, and replace readily soluble country rocks, especially carbonate ones. In this way there arise the impregnation and re-placement ores.

Hydrothermal mineral formations may be associated either with deep-seated rocks or with sub-volcanic bodies which erupted near the surface. They occur mainly as *ore and mineral veins.* The fissures in which the minerals crystallize out owe their origin to tectonic processes, and so the ore veins are confined to faults and planes of movement. Often the country rock on both sides of a mineral vein is bleached, hydrothermally decomposed or silicified, permeated by tiny little ore and mineral veins running in all direc-tions. Miners call the loamy and clay-like coatings on the lode walls 'gouge', and disturbances with shattered rocks and many small fractures are called shear zones. Ore and mineral veins occur in isolation but also as swarms, and the individual lodes of these swarms may or may not contain the same minerals. In these swarms

veins with the same parageneses often strike in one particular direction, since the joint systems opened up by tectonic forces at a given time trend in the same direction and were filled by the hydrothermal mineral solutions obtaining at that time.

The direction followed by the outcrop of the hydrothermal lodes (which are often steeply inclined) is called their *strike-direction* or strike. Ore veins often pass at depth into ore-free quartz veins; they are said to possess barren 'quartz roots'. Hydrothermal ore lodes are known in which the same minerals occur down to very great depths, while other veins are layered, with different ore and gangue minerals deposited in the various depth zones. In such cases the miner speaks of different 'levels'. The mineral content of a vein can also change in the horizontal (strike) direction, and this kind of change in paragenesis of the infilling is called a *lateral facies change*. If the individual depth zones with

◁ 60 MARCASITE greenish yellow disc-like concretion with radial acicular structure, grown in steely blue slate. From Sparta, Illinois, U.S.A. Arabic *markaschâtsâ*, pebble. FeS_2 orthorhombic. Scale 1·8:1

61 BLUE ROCK SALT (halite) translucent cubes. The colour is ▷ caused by natural radioactivity. From Stassfurt. Greek *hals*, salt. NaCl cubic. Scale 1·7:1

62 LIMONITE AND MANGANESE OXIDES brown iron dendrites and ▷ black manganese ones on a slab of very fine-grained Upper Jurassic limestone (Solnhofen beds). From Langenaltheim, near Solnhofen, Bavaria. Dendrite from Greek *dendron,* tree. Scale 2·3:1

different mineral suites follow each other quickly, one speaks of 'telescoping', and lodes (particularly vugs) with depth zones which change rapidly in this way include some classical mineral localities.

However, hydrothermal mineral solutions do not only deposit their substance in veins. Cavities in conglomerates, sandstones, tuffs and vesicular lavas may be infilled with mineral material, although no sizable crystals can be formed from the impregnation of the pores of sediments. Mineral solutions which have attacked reactive country rocks – in particular limestones and other carbonate rocks – and have replaced them with new mineral parageneses, are responsible for the *metasomatic* or replacement ore bodies. Lodes are not confined to clefts in or near the igneous body which harboured the ore solutions, for these can rise or migrate laterally and deposit their mineral content in fissures far from the parent igneous reservoir, or they can impregnate or replace alien rocks. In the case of many hydrothermal ore deposits there is no igneous rock at hand, and it can only be inferred that they owe their existence to a pluton at depth.

The minerals of hydrothermal parageneses can usually be arranged in an age sequence, since the different lode minerals have been precipitated in a definite order. Hydrothermal deposition in veins has often been repeated, each time with the same sequence of minerals. Thus we may pass twice or even more times through the succession quartz, sulphides, barytes, carbonates, and in such cases one speaks of minerals of the first, second or third

63 PYROLUSITE dark manganese dendrites standing out in relief on ▷ thinly bedded Bunter sandstone. From Ebersbach, near Aschaffenburg, Spessart. Pyrolusite from Greek *pyr,* fire, and *lusios,* destroying, since melts of green glass were bleached with this mineral. MnO_2 tetragonal. Scale 1·4:1

generation in a vein. Not infrequently earlier precipitated minerals are redissolved, and the only surviving traces of them are cavities in the minerals of later sequences. Several hundred mineral species are known from hydrothermal ore and mineral deposits, and a rich suite of *ore minerals* is comprised by compounds of sulphur, arsenic, antimony and bismuth. The gangue minerals in hydrothermal deposits are mainly quartz, calcite and other carbonates, fluorspar and barytes.

The distinguishing feature of hydrothermal mineral assemblages is the variety of mineral species found within a small area of space (half a dozen or more species are by no means rare). The variety of crystal form is outstanding, and the individual mineral species may also appear in different guise in the various generations, and may vary also in crystal habit, colour, size and surface qualities. Finally, ore minerals (sulphides, arsenides and oxides) which mostly have metallic lustre, form an attractive contrast to other lode minerals, which are of different colour and often translucent. From the paragenesis, crystal appearance and colour, and from the sequence of generations, an expert can often tell at a glance from what locality hydrothermal specimens derive, and often he can assign a given assemblage to a particular working and a particular level. Even so, the spatial arrangement of the individual crystals is different for each assemblage, and two are never identical, even if they are from the same locality.

Hydrothermal minerals are often of great economic importance and for this reason attention was turned to them at an early date. They include the most important deposits of gold, silver, copper, lead, zinc, bismuth, uranium, antimony, mercury and other important metals. Great accumulations of ore minerals are called *ore-shoots*. Such enrichment occurs mainly where veins swarm or intersect. If conversely the ore minerals in a vein are sparsely represented compared with the valueless gangue minerals, one speaks of a 'barren track'. In wide open veins the minerals are free to develop their crystal form, especially when the opening of the joint or cavity proceeds more quickly than the infilling with mineral substances. Such favourable conditions of formation allow a really lavish variety of crystal forms to grow hydrothermally, and as in the case of organisms, the development of crystal form during growth is conditioned not only internally, by the lattice type, but also by the environment. The gangue minerals form particularly fine crystals — usually the youngest ones of the whole assemblage — because they had the most ideal conditions for their growth in open passages. Extremely fine *crystal groups* have come to light as a result of ore mining at such horizons, and for this reason the names of many mines in Europe and America that have long since been disused still have a familiar ring. From many veins (long since worked out) in the Harz, the Erzgebirge, the Black Forest and other German localities unique assemblages have passed into mineral collections. And the records of many an old mine show that considerable sums were paid not only for the ores but also — even as long as two centuries ago — for show-case assemblages sold to museums or private collectors.

In contrast to the opaque ore minerals of metallic lustre, most gangue minerals are light in colour, mainly white, brown or green, some of them transparent or translucent. Many gangue minerals are readily cleaved and were accordingly named 'spars' by the miners. The *pseudomorph* minerals which occur in many hydrothermal deposits have a crystal form which does not properly belong to them. They are formed when the substance of a prior crystal is gradually taken into solution and simultaneously replaced by substance of another kind, so that the original crystal configuration is retained as a husk. Pseudomorphs are important pointers to the origin of rocks and mineral beds, and from their form the mineral species originally present can be inferred.

If the mineral solutions in a fissure can crystallize out undisturbed, the individual minerals are deposited in layers, one upon the other, on the walls, giving symmetric layers of minerals and ores. Ores growing as crusts on angular fragments of country rock broken off by tectonic movement during mineral deposition are called breccia ores (Pl. 39). The fragments of country rock with their ore coating lie embedded in a coarse mass of gangue minerals.

As already stated, the various ore minerals are deposited in a definite sequence from the mineral solutions as the temperature

◁ 64 PEA IRON ORE spherulitic limonite embedded in an argillaceous matrix. From Kandern, Württemberg. Limonite FeOOH orthorhombic. Scale 3·8:1

65 LIMONITE stalactiform group. From Gomor, Hungary. Scale 5:1 ▷

66 CHALCOPYRITE (copper pyrites) coarse brass-yellow ore, in pla- ▷ ces largely altered to dark tile ore (bottom left), a mixture of copper oxidation minerals. White vein quartz, itself veined in places by green malachite. Limonite coating at the top. From Friedrich Christian mine, Wildschapbach, Black Forest. Greek *chalcos*, copper, and *pyr*, fire. CuFeS$_2$ tetragonal. Malachite Cu$_2$[(OH)$_2$/CO$_3$] monoclinic. Scale 1·3:1

67 CHALCOPYRITE (copper pyrites) brass-yellow crystals of metallic ▷ lustre. On them are thinly tabular crystals of light-coloured calcite. Also brownspar. From St. Andreasberg, Harz. Chalcopyrite CuFeS$_2$ tetragonal. Calcite CaCO$_3$ trigonal. Brownspar (Mg, Fe, Mn) CO$_3$ trigonal. Scale 2·3:1

falls, and so the hydrothermal *ore assemblages* can be sub-divided according to their mineral content, and especially their ore content. They are also named from their characteristic minerals. Thus one speaks of the fluor-barytes lead-silver assemblage, meaning veins which bear argentiferous galena and other silver ores, with fluorspar and barytes as gangue. In these ore formations minerals containing certain chemical elements always occur together.

The veins of the gold-silver assemblage bear native metallic gold, as 'gold in the rock' grown in coarse quartz (Pl. 11), or as free gold in lamellar aggregates of different size and form (Pls. 41, 45). The greatest mass of gold in the rock (88 lb) was found in a gold-quartz vein in California. Many lodes contain argentiferous free gold and also rarer gold minerals, such as gold tellurides and selenides as well as precious silver ores. Gold-silver lodes associated with subvolcanic igneous bodies are famous for the fine crystals developed in the many vugs. Lodes containing free gold are usually ferruginous as well, i.e. coloured yellow or brown by small amounts of limonite ore. At greater depths these lodes often contain pyrite with the gold. The ore lodes of this assemblage form, with the sedimentary placer deposits, the most important occurrences of this precious metal.

The silica and copper assemblage includes important deposits where pyrite and chalcopyrite are predominant and form large ore masses. The lenses or layers are enormous, but rarely contain

68 AZURITE encrusting and reniform, intergrown with thin green ▷ needles of malachite. A cut and polished slab with some holes. French *azur*, sky-blue. $Cu_3[(OH)_2/(CO_3)_2]$ monoclinic. Malachite $Cu_2[(OH)_2/CO_3]$ monoclinic. Scale 1·2:1

fine crystals. Other representatives of this ore assemblage are the impregnation deposits, likewise with considerable accumulations of copper and other metals, but without large-sized crystals. Fine mineral groups occur in this ore assemblage only when the mineral substance has been deposited in ore-veins or in the vesicles of igneous rocks. Native copper occurs in the vesicles of many old basalts, where it forms coarse metallic masses of up to several tons. Many-branched aggregates of distorted crystals of strange form also occur (Pl. 40), and also lamellar forms as a kind of natural metal foil.

An important mineral in this assemblage, and also the most abundant copper mineral, is chalcopyrite, which usually forms compact masses (Pl. 66). Well-developed, brass-yellow crystals are not common (Pl. 67), and when they do occur they are often twinned and have their many faces distorted. Like other copper sulphides chalcopyrite often has a tarnished surface layer of bright colours.

Most of the earth's hydrothermal ore veins belong to the lead-silver-zinc assemblage. Argentiferous galena, zincblende, pyrite, tetrahedrite, chalcopyrite, bornite and silver ores are the principal ore minerals of this suite, and many lodes are known which bear fine groups. In many veins vugs with crystals of galena (Pls. 3, 43) and zincblende (Pl. 44) are as plentiful as complicated crystal forms of the gangue minerals, in particular calcite (Pls. 13, 14). *Galena* often contains small amounts of silver and so becomes a valuable silver ore. In addition to the predominant cubic crystals it also occurs in combinations with many faces, and sometimes as individuals of considerable size. Skeletal, compact, nodular and granular aggregates are found. Galena which has been rolled and elongated by tectonic movement is known as foliated galena. The mineral is the same colour as metallic lead, and on fresh cleavage surfaces it has a strong metallic lustre. Its surface is often tarnished. The mediaeval miners called minerals 'blendes' when they looked like metallic ores, although the metallurgy of those days could not extract the metal from them. The name *zinc* comes from the often jagged looking crusts, the 'zincs', which on smelting are deposited in the ovens and are sometimes known as 'stove-calamine'.

Zincblende (Pl. 48) – zinc sulphide mixed with iron, cadmium, manganese, and other elements – forms cubic crystals, often distorted and with many faces. In addition zincblende also occurs as twinned crystals, and as irregular grains. The crystals have a strongly adamantine lustre, are opaque and coloured dark brown to black, although they may also be of lighter colour (yellow, reddish or brown) and translucent. The rhombohedral cleavage planes have adamantine lustre. Yellow and light-brown translucent zincblende is called honey blende, red crystals are 'ruby blende'. Sometimes the mineral occurs as layers of reniform, encrusting finely fibrous or dense aggregates, which often include wurtzite — another polymorph of zinc sulphide — and galena (Pl. 51).

In the lead-silver-zinc association and in other hydrothermal ore deposits we meet *grey copper* ores (tetrahedrite), distributed both as compact masses and as tetrahedral crystals with many faces, which are often striated, and have ideally sharp edges. The most important members of this grey copper ore family contain copper and silver, sometimes also zinc, iron or mercury, and in addition sulphur, antimony, arsenic or bismuth. Thus a whole series of individual mineral species may be present in one of these assemblages (Pl. 52). Another assemblage is characterised by the association of minerals of the elements *silver, cobalt, nickel, bismuth* and *uranium,* and the veins of this kind contain particularly large numbers of mineral species – two dozen or more in a single vein are not uncommon. Examples of well crystallized minerals from this association are silver (Pl. 46), ruby silver

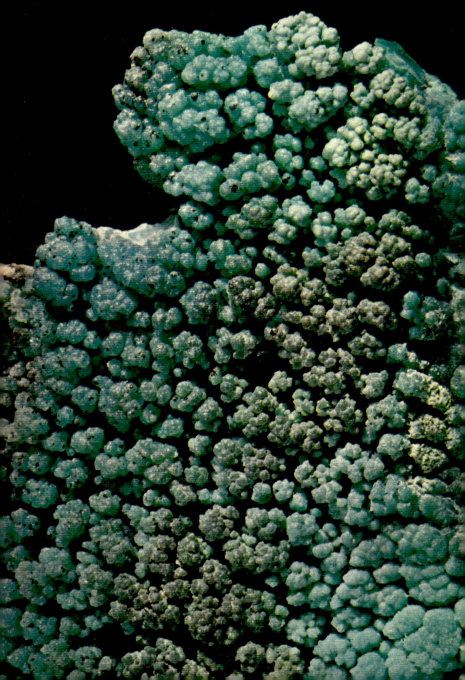

(Pl. 47) and carbonate gangue minerals (Pl. 7). Dendritic aggregates of native silver are often built of very small cubes and octahedra (Pl. 46). The red translucent ruby silver (Pl. 47), which occurs in large crystals of adamantine lustre and in fine assemblages, is included by miners among the valuable silver ores. In the old days cloths were laid under workings of ruby silver, native silver and other highgrade silver ores so that none of the precious material should be lost. Most ruby silver ores are sensitive to light, particularly to sunlight, which deprives them of their bright lustre and tarnishes them with a grey film of silver. Hence these valuable ore assemblages must be protected from strong daylight in mineral collections.

From even cooler mineral solutions – below the boiling point of water – the ores of the *antimony-mercury association* are precipitated. This formation is of comparatively rare occurrence and has a more limited range of minerals than the others already described. The most important antimony mineral of this kind of occurrence – in fact the most important antimony mineral altogether – is antimonite, with its abundance of forms. Miners used to call this grey mineral, with its bright lustre and spear-like crystals, grey antimony glance (Pl. 85). A valuable mercury ore from this formation is cinnabar.

◁ 69 CHRYSOCOLLA green warty crystals of gelatinous origin. From Lipari islands, Italy. Greek *chrysos*, gold, and *kolla*, glue, because the mineral was used for soldering gold. $CuSiO_3 \cdot nH_2O$ cryptocrystalline. Scale 3·4:1

70 WAVELLITE thin disc-like radial acicular crystals formed on ▷
71 blue-grey siliceous slate. From Langenstriegis near Freiberg, Saxony. After Wavell who discovered it. $Al_3[(OH)_3/(PO_4)_2] \cdot 5H_2O$ orthorhombic. Scale 4·3:1

Mineral veins of the oxide *iron-manganese-magnesia association* are widely distributed. They arose from relatively low-temperature solutions and usually contain only a few mineral species. The most important are spathic iron ore (siderite), haematite, manganese minerals of the brownstone group, manganese spar and magnesite. Whereas in the ore formations previously described the characteristic minerals are sulphides, here oxides, hydroxides and carbonates are predominant. Siderite occurs in veins, and as a replacement body in large stock-like masses. Magnesite also forms extensive replacement bodies in limestones and dolomites. In iron-manganese ore lodes there are fine crystal groups and aggregates of rhodochrosite, the raspberry-red manganese carbonate mineral (Pls. 4, 57).

Finally there is an ore-free hydrothermal vein assemblage – veins filled exclusively with gangue. They contain quartz, fluorspar, barytes, calcite and other minerals. Vugs in these veins are excellent localities for perfect many-faced crystals of barytes and fluorspar, the former occurring in many forms: thick tabular or prismatic crystals with chisel-like ends; thin tabular or lamellar ones, or rosettes and fan-like groups; lamellar and scaly aggre-

72 CHALCANTHITE (copper vitriol) group of blue triclinic crystals ▷ of tabular habit, with needles of white iron vitriol (melanterite). From Coquimbo, Chile. Vitriol, mediaeval Latin *vitriolum*, alchemic term for soluble sulphates, from Latin *vitrens*, glassy. Chalcanthite from Greek *chalcos*, copper, and *anthos*, bloom, because of the bloomed colours of copper ores. $Cu[SO_4] \cdot 5H_2O$ triclinic. Melanterite from Greek *melanteria*, because it yields a black colour with tannic acid. $Fe[SO_4] \cdot 7H_2O$ monoclinic. Scale 2·6:1

gates, and compact masses (Pls. 6, 53–4, 59). Fluorspar, with its almost moist-looking vitreous lustre occurs mainly as cubes and only rarely in combinations with a larger number of faces. It can be colourless and as clear as lens glass, or have any one of a large number of colours (Pls. 8, 53–4, 59).

In addition to calcite in various habits and many different shades of colour, other carbonates occur in hydrothermal fissures as gangue minerals. In addition to dolomite (Pl. 7) and siderite (Pl. 43), ankerite or brownspar (Pl. 67) is widely distributed. Dolomite, siderite and brownspar normally have considerably fewer faces than the calcite crystals. Light-coloured siderite assumes a dark brown colour when it has been weathered, as on mine tip-heaps.

Quartz is almost always present in the hydrothermal veins of all ore assemblages – not only in association with other minerals, but also as almost completely pure quartz veins. Even veins which contain a variety of minerals at high levels may contain only quartz in depth. The quartz appears in many forms. When it is massive and compact it is coarsely crystalline and without vugs. It may be clouded milky by very fine liquid inclusions (milky quartz). Vein quartz is often dense like hornstone, layered like chalcedony and present in several generations. Quartz can pseudomorph a number of other vein minerals. When thinly tabular barytes is replaced by quartz, what miners call 'hacked quartz' is formed. Well-shaped quartz crystals often project into the cavities of hydrothermal veins, and the crystals of the youngest quartz generation are often perfectly transparent.

As a result of erosion magmatic and metamorphic rocks appear at the surface and are exposed to weathering. Disintegration and decomposition of their minerals then proceeds quickly or slowly according to the climate. New minerals and rocks may be formed from the debris and from the material taken into solution, and rocks formed from surface deposits in lakes, rivers and the sea are called *sediments*.

Mechanical weathering is facilitated by slaty-cleavage, jointing and fissuring, and so heaps of blocks, detrital gravel or fine-grained debris are formed, all of which are transported by gravity, water, wind, ice or marine currents, and deposited to form clastic or *detrital rocks*. Such sediments are initially unconsolidated and are cemented only in the course of geological time. The mineral grains and rock fragments in the detrital rocks allow us to infer the mode of transport. Thus grains and pebbles transported by wind and water are typically rounded; but although some larger boulders and pebbles may be striated when they are transported by ice, smaller fragments are characteristically angular and moraine deposits contain completely unsorted mixtures of large boulders and small fragments. Consolidated sediments of rounded pebbles are *conglomerates;* those made of angular fragments are *breccias*. Entirely different kinds of sediments arise from precipitation in evaporating basins or in freshwater lakes. Sea deposits are called *marine*, freshwater lakes give *limnic* sediments. Materials accumulated as river deposits are fluviatile and deposits from glaciers are glacial, while melt-water deposition gives *fluvioglacial* sediments. *Terrestrial* beds form on dry continents, *aeolian* ones are due to wind transport. Most sediments are formed by

the various inorganic processes outlined above, but in addition organisms can contribute to their formation in a variety of ways.

Formation of sediments depends on many conditions and *sedimentary rock types* vary according to the initial material (from which the clastic deposits are formed), the manner and duration of the transport, the nature of the environment of deposition and also the climate. *Diagenesis* is the consolidation which affects loose sediments in the course of time, and is achieved by compaction, and by permeation with lime, silica or other mineral substances in solution. It is not always the case that particularly old sediments are well cemented or even compacted, and clays of great age are known (e.g. the Cambrian blue clay of Leningrad) which are still soft and plastic, whereas much younger gravels have been cemented to hard conglomerates. According to their mode of origin sedimentary rocks have distinctive characters — minerals, grain-size of the mineral species, texture – and these

◁ 73 MALACHITE fibrous clusters of elongated prism needles, with earthy red copper ore (cuprite) below. Yellowish calcite. From Gumishev, Urals. See Pl. 9–10. Cuprite Cu_2O cubic. Scale 3·1:1

74 PHOSPHOCHALCITE (phosphatic copper ore) encrusting and ▷ nodular at the margins, thin dark green crystal needles in the centre. The base is coarse milky white quartz. From Virneberg, Siebengebirge, Germany. Greek *phosphoros,* light-bearer, and *chalcos,* copper. $Cu_5[(OH)_4/PO_4]_2$ monoclinic. Scale 1·6:1

75 MALACHITE green prismatic crystal and smaller stumpy tabular ▷ crystals of blue azurite. From Tsumeb mine, S.W. Africa. See Pl. 9–10. Azurite $Cu_3[(OH)_2/(CO_3)_2]$ monoclinic. Scale 5·2 :1

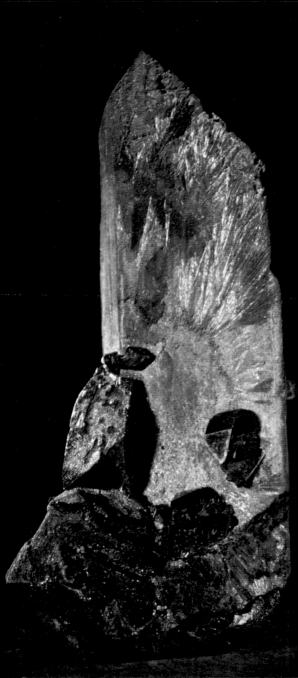

characteristics constitute their *facies*. The crust of the earth is almost everywhere covered by a blanket of sediments that varies very much in thickness, and so sedimentary petrography (the study of the origin and nature of sediments) is an important branch of geology.

At the earth's surface the magmatic and metamorphic minerals, which were formed under temperature and pressure conditions very different from those obtaining there, are differentially broken up or dissolved. Their commonest minerals, the felspars, are particularly sensitive and weather to form clay minerals. Quartz on the other hand, being very hard and without cleavage, is not easily broken down mechanically. It is also chemically very stable and practically insoluble, and for these reasons quartz is the most widely distributed mineral on the earth's surface and the most important component of many clastic sediments. As loose sands and as consolidated sandstone it forms enormous rock masses. When the dark micas, the *biotites,* weather, their iron content is first dissolved out and they acquire a golden lustre (so called 'golden mica'). If weathering proceeds further they lose their magnesium and become silvery bright ('argentine mica'). The weathering of ferruginous minerals produces iron hydroxide, which occurs in many sediments as limonite and colours them brown or red. Rocks consisting predominantly of limonite are valuable iron ores. Limonite also exists in the modifications of goethite and micaceous goethite. Weathering solutions containing compounds of iron and manganese — oxides and hydroxides — are

76 CERUSSITE (white lead ore) white stalky and spear-shaped ▷ crystals, some coated with blue azurite. From Claustal, Harz. Latin *cerussa*, white lead. $PbCO_3$ orthorhombic. Scale 3:1

constantly being supplied to sediments, and they give rise both to limonite and to various aggregates of manganese minerals. In fissures the mineral content of such solutions can be deposited as dendrites (Pls. 62, 63).

Among the minerals formed in sedimentary rocks the *clay minerals* are particularly important as rock-builders. These exceedingly minute mineral flakes derive from the weathering of felspars as noted above, and from other aluminium silicates. In a warm and wet climate potash felspar, for example, is converted into kaolinite. Clay minerals are incapable of forming crystals of great size, and their individual mineral particles can only be seen under high magnification. Many other sedimentary minerals, particularly if they are the products of chemical decomposition, form powdery, earthy and often loose fine-grained masses.

Calcite, which as limestone forms large rock masses, is the commonest carbonate in sediments. *Dolomite* holds second place, and like calcite it may form thick deposits which build whole mountain masses. Organisms often contribute to the formation of limestones and dolomites, as also to phosphatic and siliceous rocks. Peat, lignite, coal and anthracite are economically important sedimentary rocks originating from plants.

A characteristic common to most sedimentary rocks is their *bedding*. It is particularly striking where deposition occurred slowly and in calm conditions, so that the sediments were deposited in layers, one above the other, each layer being bounded by more or less parallel planes, the bedding planes. Thinly bedded sediments may be described as laminated. Thicker sequences of sedimentary rocks of different types form a *series of strata* or a stratigraphic series. The beds are said to be concordant when younger beds follow older ones without interruption and with parallel stratification. On the other hand one speaks of unconformities when sedimentation was interrupted for some considerable period, or when younger beds lie on much older ones which have been folded, e. g. on the tectonically disturbed roots of ancient mountains. In addition to parallel bedding, which can be seen very clearly in marine deposits, there are also *cross-bedded* and *current-bedded* sediments, which are formed in rivers and lakes or on shores with fluctuating currents or, in the case of aeolian sediments, when the wind is subject to changes in direction.

Concentrations of valuable minerals concordantly bedded in the sedimentary sequence are known as *seams*. The top of a seam is its hanging-wall, the bottom is its foot-wall. Like other beds seams may become thinner when traced laterally – they 'wedge out'.

Finally the sedimentary rocks, particularly those of marine origin, may also contain *fossils,* and from these remains of animal or plant organisms the evolution of life on our planet can be traced. In many cases only the hard parts of these prehistoric organisms are preserved, petrified in limestone or by means of quartz or other mineral substance. Collectors prize the fossil fish from the Kupferschiefer, preserved in chalcopyrite, zincblende, silver ores or native silver, and also the pyritized ammonites ('golden snails') from the Lias. Wood agate derives from silicified wood, and from many geological formations whole silicified forests are known. Among the minerals of organic origin is amber from the resin of prehistoric conifers. In the yellow-brown transparent nodules of this amorphous mineral one finds inclusions of conifer leaves, insects and other small organisms.

Many magmatic and metamorphic minerals are very resistant to weathering, e. g. garnet, which is difficult to break up mechanically because of its hardness and lack of cleavage. Often it is the minerals of high specific gravity which survive the weathering and solution of rock masses. These *heavy minerals* can be accu-

mulated by the winnowing action of running water, wind or marine currents, as the lighter mineral material is washed or blown away. Accumulations of resistant or unweatherable minerals which survive the destruction of their parent rocks are called *placers*. Placers of noble metals contain gold or platinum, while other placers consist of rolled grains of valuable ore minerals. Thus placer tin is distinguished from mountain tin – the tin in pneumatolytic parageneses – and placer gold from 'gold in the rock'. Gold in river placers exists as very fine leaf-gold (like the Rhine gold of the legends) or as nuggets of cherry-stone size, or even as heavier rounded bodies. The largest placer-gold nugget so far discovered weighed 154 lbs. Well-formed crystals are not to be expected in placer deposits. However, many gem-placers have been known for centuries as collecting localities for large, particularly clear and naturally sorted crystals, for only minerals which are free of fractures and inclusions can survive the buffeting of transport with other pebbles along a river bed. Among the gems obtained from placers are diamond, zircon, sapphire (Pl. 92), ruby, topaz, tourmaline and garnet.

◁ 77 TORBERNITE lamellar flaky aggregate of copper uranite (uranmica) consisting of thin tabular crystals. The base mineral is barytes. From Vernon, Cornwall. After the Swedish chemist Torbern Bergman. $Cu[UO_2/PO_4]_2 \cdot 10H_2O$ tetragonal. Scale 2·2:1

78 PYROMORPHITE (green lead ore) grass-green prismatic crystals ▷
79 on limonite, projecting into a vug. From Schauinsland, Black Forest. Greek *pyr*, fire, and *morphe*, form, because it changes shape in the soldering tube. $Pb_5[Cl/(PO_4)_3]$ hexagonal. Scale 2·3:1

Sedimentary rocks make but a small contribution to the totality of the solid crust, yet they are widely distributed at the surface both as loose and as consolidated sediments. Fine mineral assemblages are nevertheless rarely found in normal sediments, partly because suitable cavities in which sizable crystals can grow are exceptional. However, in fine-grained mud, clay and other soft sediments quite large crystals of pyrite, chalcopyrite, or gypsum can form because of the plasticity of the host rock. Well-developed crystals also occur in the hollow interior of fossils, and several mineral species are known as agencies of fossilisation. In clay or marl *marcasite* forms concretions of many shapes. Concretions are accumulations of mineral substance in the form of spherical and irregular nodules or radial acicular aggregates which have grown outwards from a centre. Marcasite concretions formed in clay, in de-oxygenated mud or carbonaceous sediment, often become converted to pyrite in the course of time.

Elementary *sulphur* is formed in nature in a number of ways: in volcanic craters; as a deposit from sulphurous hot springs; from the activity of sulphur bacteria; and by reduction of sulphates (such as gypsum or anhydrite), effected by organic substances. Sulphur is also formed in small amounts in the oxidation zone of sulphide ore deposits. The reduction of sedimentary sulphate rocks has not only produced economically important sulphur deposits but also the finest and largest sulphur crystals (Pl. 5). In these conditions sulphur commonly occurs in association with celestine (Pl. 58), calcite and aragonite.

80 CROCOITE yellowish-red thin elongated prisms. From Dundas ▷ area, Tasmania. Greek *krokos,* saffron, because of its orange-red streak. Pb[CrO₄] monoclinic. Scale 2·5:1

Among the chemically precipitated minerals are *salts* of various kinds, forming extensive deposits. Marine salt beds were deposited from sea-water which was evaporated by the heat of the sun in basins with restricted access from the open sea. As a result, the salts dissolved in sea-water (mainly rock salt) were precipitated. Salts and salt sequences of a quite different kind (in the form of fine efflorescences and coatings, with a larger number of minerals) arise from evaporation in inland lakes with no drainage outlets.

During the earth's history enormous quantities of salts have been taken into solution as a result of rock weathering and carried into the sea by river transport. Further salt has been formed by volcanic action, and so the salt content of the oceans has steadily risen, for only a small amount has been re-extracted by the formation of marine salt deposits. If all the world's oceans were evaporated and the salts precipitated spread evenly over the ocean beds and continents, then our planet would be covered with an evaporite layer 115 ft thick. The most important and the commonest evaporite is *rock salt,* the only mineral suitable for human consumption in its natural state. It crystallizes in cubes and is occasionally coloured blue by radioactive agencies (Pls. 16, 61). Cooking salt is obtained by mining salt-springs and solid salt deposits, and by evaporating sea-water in salt pans. Many evaporite minerals are stable only in a dry atmosphere and deliquesce in a damp environment. In mineral collections these hygroscopic minerals must therefore be kept out of contact with moist air.

6 MINERALS IN THE ZONES OF OXIDATION AND CEMENTATION OF SULPHIDE ORE DEPOSITS

When ore deposits containing pyrite and other ferruginous sulphides are weathered, limonite is formed just below the ground surface. Miners used to call limonite masses formed in this way *iron cap*. The term has been extended to cover the oxidation zones of lodes with different mineral content, so that one speaks quite generally of the *cap zones* of sulphide ores, whether limonite is present or not.

In the zone of weathering minerals are decomposed and taken into solution, while new minerals are precipitated. In the upper regions of this zone, where oxygen is plentiful, the sulphide and arsenide ores are oxidised and dissolved, and percolate down to levels below the water-table where they can be precipitated in a *zone of cementation* – a region often greatly enriched by accumulations of quite noble metals such as copper or silver. Often gold and newly-formed sulphides are found here. As a result of processes of this kind oxidation zones often preserve enormous quantities of metal which derive from higher horizons which have long been removed by erosion. The ore content of many unweathered mineral veins is insufficient for economic exploitation, while the cementation zone may yield unusually large ore quantities with a high proportion of metal.

Since the least stable minerals tend to weather out, the veins in the oxidation zone normally appear cellular, spongy, or carious, with well-formed crystals of new oxidation minerals in the cavities. Particularly thick cap zones are formed in a warm dry climate, where the solutions of weathered minerals are not quickly diluted and removed by rainwater. In areas of heavy rainfall the

cap zones above the ore beds may be extensively leached and in this way greatly depleted of oxidation minerals. The character of the country rock is an important factor. Highly reactive acid mineral solutions precipitated in limestones or other carbonate rocks may form strikingly beautiful oxidation minerals. Other phenomena occurring in the oxidation zone are the migration and redeposition of mineral material, and the impregnation of the country rock with oxidation minerals.

Many kinds of *pseudomorphs* are formed in sedimentary rocks, particularly in cap zones. They have retained the form of a crystal which existed on the site before them, but have replaced its substance. Because of their excellent crystal form they are much sought after by collectors. Various kinds of pseudomorph may be distinguished. A *replacement pseudomorph* is formed when the original crystal substance is removed by leaching and at the same

◁ 81 WULFENITE honey-yellow thin tabular crystals of resinous lustre on weathered galena. From Bleiberg, near Villach, Austria. After the Austrian mineralogist Wulfen. $Pb[MoO_4]$ tetragonal. Scale 2·5:1

82 TURQUOISE bluish reniform aggregate of tiny triclinic crystals ▷ with waxy lustre. From Nischapur, Chorassan, Persia. After the Chaldee word for this stone, *torkeja*. The mineral reached Europe via Turkey. $CuAl_6[(OH)_8/(PO_4)_4] \cdot 4H_2O$ triclinic. Scale 2·8:1

83 SMITHSONITE nodular whitish crusts on dark altered zincblende. ▷ From Nertshinsk in Transbaikalia, U.S.S.R. After the English chemist Smithson. $ZnCO_3$ trigonal. Scale 1·7:1

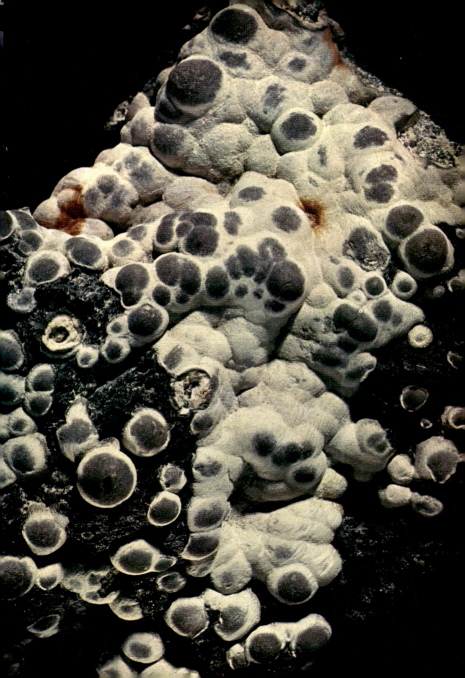

time replaced by mineral substance of another kind. But if the original substance is chemically changed into another, again without impairing its crystal form, one speaks of a *transformation pseudomorph*. This mineral reconstitution may be effected by partial removal of the original substance and partial replacement with new material (not necessarily occurring simultaneously), leaving some of the original material intact. Finally, a crystal may be enclosed completely by another mineral species formed as a crust round it, leaving the form of the original recognisable. If the original mineral is then dissolved, a negative pseudomorph, a *perimorph*, is formed – a hollow preserving the crystal form of the original mineral.

Colloidal minerals are typically developed in oxidation zones. They are globular, nodular masses formed from glutinous mineral gels, and often owe their existence to oxygen-rich surface waters. They become crystalline in time, and can be recognised from their lustrous rounded, globular, reniform or botryoidal surface (Pls. 65, 69).

The minerals of the oxidation zone are mainly oxides, hydroxides or oxygen-rich salts of heavy metals – carbonates, sulphates, phosphates, arsenates, molybdates, chromates, wolframates, vanadates and others similarly rich in oxygen. The *oxidation minerals* do not only occur in many different forms of aggregates but also as almost ideally developed individual crystals. Depending on such factors as climate, the original ore minerals, and the nature

84 ADAMITE blue-green warty crystal aggregates on earthy limonite ▷ in cavities in yellowish calcite. From Laurion, Attica, Greece. After the French mineralogist Adam. $Zn_2[OH/AsO_4]$ orthorhombic. Scale 2·2:1

of the country rocks, the aggregates may be multi-coloured and give rise to granular, radiating, lamellar, layered, encrusting or reniform masses; or they may form only thin bright-coloured films. Powdery and earthy masses formed by the decomposition of heavy mineral ores have been named ochres. Antimony ochre (Pl. 85) for example is a mixture of powdery, earthy or encrusting minerals formed from the decomposition of antimony ores such as hydroromeite, stibioconite and cervantite. Hydrated metallic sulphates are named vitriols, and they include copper vitriol and iron vitriol (Pl. 72), common minerals in the oxidation zones of warm and dry climates. The iron cap of sulphide ore deposits harbours many mineral parageneses which are often strikingly rich in colour, and it was in many cases these brightly coloured cap minerals which drew attention to the valuable ore-accumulations hidden in the zones below. However, oxidation ores are easily smelted, and so they were sometimes the ones valued most. In many instances they alone could be exploited, since the sulphide ores below them did not respond to primitive metallurgical techniques. Thus even in antiquity smithsonite and other zinc carbonates and silicates were worked as ores, while zincblende, which weathers to form these cap minerals, has only been smelted in any quantity during the past hundred years.

Often the *colour of the cap ores* gives an indication of the unweathered minerals lying below. Particularly bright and gaudy coloured oxidation minerals result from the decomposition of pitchblende and other uranium minerals. Blue and green predominate among oxidation minerals containing copper. These include green malachite (Pls. 9–10, 73, 75), green chrysocolla (Pl. 69), green phosphochalcite (Pl. 74), blue azurite (Pl. 68) and blue copper vitriol (Pl. 72). Malachite is formed from the weathering of various copper ores, especially chalcopyrite. It forms thin fissure infillings and coats the country rock green, thereby suggesting a much richer copper deposit than may in fact be there. Malachite may be accumulated as enormous masses, and in the Urals blocks of it weighing several tons have been found. If the original minerals contain arsenic, as with tetrahedrite or enargite, then azurite usually forms in the oxidation zone. Like malachite, dark blue azurite is a hydrated copper carbonate and – again like malachite – occurs not only as well developed crystals but also as encrusting, reniform and botryoidal masses of radial acicular structure (Pl. 68). Brown or blackish-brown earthy mixtures of copper oxidation minerals – especially of cuprite (red copper oxide) occurring with limonite – are called tile ore (Pl. 66). Phosphochalcite (also known as pseudomalachite) owes its origin to the action of phosphate solutions on copper ores which are being weathered. It is pistachio-green and forms encrusting and downy aggregates which consist of slender individual crystal needles (Pl. 74). The green mineral chrysocolla (a hydrated copper silicate, Pl. 69) is formed – like the much rarer emerald copper, dioptase, sometimes used as a gem-stone – from siliceous weathering solutions. The grass-green mineral torbernite (Pl. 77) also contains copper. It is a hydrated phosphate of copper and uranium, known as 'uran-mica' because of its micaceous cleavage. Finally adamite (a zinc arsenate) and turquoise (hydrous phosphate of aluminium) owe their green colour to their copper content. They are depicted in Pls. 84 and 82.

Lead is present in a great many brightly coloured oxidation minerals which form good crystals. Miners speak of blue lead, red lead, green lead, white lead ore, horn lead, and lead vitriol. A number of these minerals are found in lead lodes which have been hydrothermally or metasomatically reconstituted, or as infillings of cavities in galena which reactive solutions have eaten into, where they form excellent crystals. Pyromorphite or green lead ore often coats extensive surfaces with crystal mats – groups

of barrel-shaped, prismatic, spear-shaped or acicular crystals (Pl. 78–9). Wulfenite crystals occur as stumpy prisms and pyramids or form tabular aggregates. From its colour (wax-yellow or orange-yellow) it is known as yellow lead ore (Pl. 81). Cerussite or white lead ore (Pl. 76) is colourless, grey or white. Crocoite or red lead ore, a lead chromate, is built from elements which derive from entirely different geochemical provinces: chromium belongs to early orthomagmatic crystallization, while it is the hydrothermal residual solutions that are enriched in lead. Crocoite (Pl. 80) is therefore rare and can occur only when weathering solutions of galena or other lead ores are brought into contact with country rocks containing chromium.

Quite a large number of oxidation minerals derive from the weathering of zinc ores, particularly zincblende. When weathering solutions containing the zinc ion react with carbonate country

◁ 85 ANTIMONY OCHRE yellow crusts on spear-shaped crystals of metallic looking antimony glance (stibnite). Quartz crystals on the right. From Felsobanqa, S.E. Carpathians, Hungary. Latin *antimonium,* from Arabic *al-ithmidum,* metallic lustre. Stibnite from Latin *stibium,* steely lustre. Stibnite Sb_2S_3 orthorhombic. Antimony ochre, a mixture of antimony oxide and hydroxide. Ochre from Greek *ochra,* pallor. Scale 2·3:1

86 ANDALUSITE a cross-section of reddish-white crystal of the ▷
87 variety known as chiastolite with dark carbonaceous inclusions in the form of a cross. From Gefrees, Fichtelgebirge, Germany. So named from its occurrence in Andalusia. Chiastolite because of its cruciform shape resembling the Greek letter X. $Al_2[O/SiO_4]$ orthorhombic. Scale 7·4:1

rocks, smithsonite (Pl. 83) is formed, sometimes as small grey crystals, often closely packed, sometimes as encrusting layered and nodular masses which may be coloured by impurities. Mixtures of smithsonite with other zinc oxidation minerals (e.g. zinc silicate or hydrozincite) are called calamine. Adamite may be formed from the action of arsenious solutions on zinc minerals, and exists as finely crystalline encrusting aggregates, or as blue-green crystals of tabular habit and vitreous lustre which may be combined into warty groups (Pl. 84). Garnierite is a nickel silicate, formed not from the weathering of nickel minerals in fissures, but from nickeliferous rocks rich in olivine. It forms large, nodular, encrusting and massive aggregates, green in colour and finely fibrous in texture (Pl. 88).

The most important and the commonest mineral of the cap zone is limonite (Pl. 65). Although it does occur as a deposit from mineral solutions which rise up from depth, it is more often deposited by surface waters as they percolate down. Another possibility is that it is precipitated when descending and uprising solutions meet. In the cap zones limonite undergoes extensive redistribution by solution and redeposition. Because it can be formed in so many ways it occurs in a great many forms which have been separately named, although they consist of the two crystal species goethite and micaceous goethite.

Although many oxidation minerals are colourful enough, their colours are often short-lived, since they deliquesce in a moist

88 GARNIERITE apple-green finely crystalline mass with reniform ▷ surface, coated with limonite. From Numea, New Caledonia, Pacific. After the Frenchman Garnier who discovered it. $(Ni, Mg)_6[(OH)_8/Si_4O_{10}]$ monoclinic. Scale 1·7:1

atmosphere. This is particularly true of minerals formed from water drops which drip steadily down in the worked-out ore horizons of mines or in disused pits. These minerals, most of which are stalactiform, are preserved in all their beauty in mineral collections by being kept out of contact with the air. It is in the oxidation and cementation zones of sulphide deposits that the majority of ores are formed. And the quantities produced in this way depend not only on the minerals available for weathering, but also on such geographical factors as climate (in particular the amount and seasonal distribution of rainfall), the level of the water-table, and of course the nature of the country rock. How rich a sulphide ore horizon is in oxidation and cementation minerals does not depend on the number of the vein minerals, nor on the manner in which they occur. Many lodes which bear few primary ore minerals, with poorly-developed crystals, are nevertheless famous because of the splendid deposits in their cap zones.

Metamorphism, the transformation of minerals and rocks, may occur when temperature or pressure is increased at depth. Metamorphic processes are usually correlated with the uprise of magmatic liquids or with mountain-building movements. Both sediments and igneous rocks can be affected by metamorphism, and rocks which are already metamorphic can be metamorphosed more intensively. Those metamorphic rocks which derive from igneous ones are designated with the prefix 'ortho', while 'para' is used as a prefix when referring to metamorphic rocks of sedimentary origin. In this sense orthogneisses can be distinguished from paragneisses.

The minerals in a rock undergoing metamorphism may be preserved, as when limestone is converted to marble: the calcite crystals of the original rock are simply *recrystallized*. But it more often happens that metamorphic processes not only coarsen the grain size of the crystals, but also lead to the formation of minerals which were absent from the parent rock. Metamorphic changes may leave the overall chemical composition of a rock unchanged, but it is just as common for mineral material to be brought in or taken out — transported by the water that permeates the rock fabric or forms a thin film along the boundaries of the grains. When rocks are under mechanical stress, the minerals first yield by plastic flow and then fracture. *Cataclasis* has then occurred; new minerals may be formed in such rocks by recrystallization, and fissures filled with calcite, serpentine, quartz or other minerals. If a rock is subjected to even stronger mechanical stress, so that it is brecciated, and if the crushed and pulverised fragments are recemented by recrystallization or by the supply of new mineral material, then a *mylonite* is formed.

Regional metamorphism is the term used to designate metamorphic changes that have affected a large area. *Dynamometamorphism* occurs in orogenic regions in association with folding and overthrusting. Even the increase in pressure caused by burial under sedimentary or other rocks thousands of metres thick is sufficient to cause some metamorphic changes.

The different temperature and pressure conditions under which rock transformations occur allow *grades* of intensity of metamorphism to be distinguished. Of lowest grade are the processes in the *epizone,* where minerals with a considerable water content can

◁ 89 RHODONITE rose-red with cracks infilled with black oxides of manganese. Cut and polished slab. From Happy Camp, Siskiyou Country, California, U.S.A. Greek *rhodon,* rose. $(Mn, Fe, Ca)[SiO_3]$ triclinic. Scale 1·1:1

90 LAPIS LAZULI coarse dark blue grains embedded in white marble, ▷ with yellow crystals of pyrite. From Malaja Bistraja, Lake Baikal, U.S.S.R. Lapis lazuli from Latin *lapis,* stone, and Arabic *azul,* blue. $(Na, Ca)_8[(SO_4, S, Cl)_2/(AlSiO_4)_6]$ cubic. Pyrite from Greek *pyr,* fire, because it gives off sparks when struck. FeS_2 cubic. Scale 1·1:1

91 CLINOCHLORE thick 'books' (some of them bent) of lamellar ▷ crystals, on which lie hyacinth red translucent calc-aluminium garnets (hessonite). From Mussa Alp, Ala Valley, Piedmont, Italy. Greek *klino,* incline, because of its crystal form; and Greek *chloros,* greenish yellow. $Mg_5Al[(OH)_8/AlSi_3O_{10}]$ monoclinic. Garnet from the colour of the pomegranate. Hessonite from Greek *hesson,* less, because less valuable than true hyancinth (i.e. zircon). $Ca_3Al_2[SiO_4]_3$ cubic. Scale 2·2:1

still survive. An example of metamorphism in this zone is the conversion of slate into phyllite — a still cleavable rock differing from slate in having mica flakes big enough to see and possessing a silky lustre. It consists principally of newly formed minerals such as chlorite, sericite, albite or zoisite, which derive from the clay minerals originally present. When phyllites are exposed to more intensive metamorphism, in the *mesozone*, the mineral suites of the epizone are in turn reconstituted, giving mica schist. Chlorite is changed to biotite, fine-grained sericite to the coarser form of muscovite. The mesozonal mica schists resemble the epizonal phyllites in that they too contain an abundance of mineral flakes whose parallel orientation gives a schistose texture to the rock. Well-developed crystals sometimes grow in these schists. Examples are garnet, magnetite, kyanite, dolomite, pyrite and staurolite, the last-named forming distinctive interpenetrant twins (Pls. 95, 96).

In the *katazone* rocks are exposed to even higher temperatures, and the mesozonal minerals give way to new assemblages which typically include felspar, biotite, pyroxene, amphibole, garnet spinel, cordierite and sillimanite. The rocks characteristic of the katazone are the gneisses, in which segregation of the different minerals into distinct and contrasting laminae is more conspicuous than the degree of parallel orientation of the minerals, i.e. the schistosity. The epi-, meso- and katazones represent a progressive increase in intensity of metamorphism, and as intensity is nor-

92 SAPPHIRE rolled blue corundum crystal of barrel-shaped habit ▷ from a placer deposit. From Ceylon. Sapphire from the Arabic, said to be derived from the island Sapphirine in the Arabian Sea. Corundum from Sanscrit *kuruwinda*, name of a jewel. Al_2O_3 trigonal. The blue colour is due to the presence of iron and titanium. Scale 1·2 :1

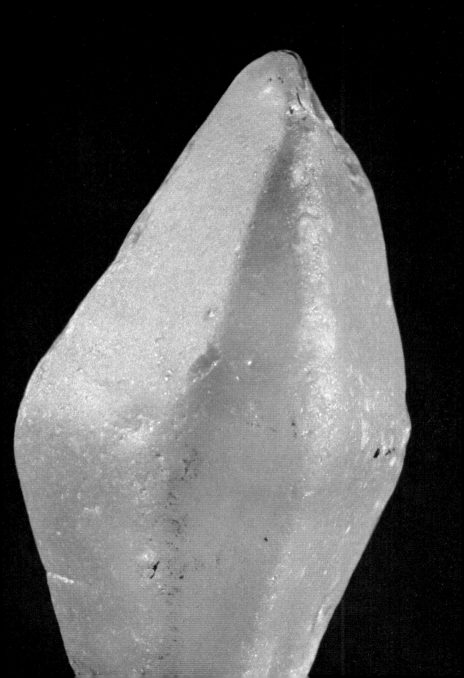

mally correlated with depth, these zones correspond on the whole to depth zones within the earth's crust.

Parallel texture, giving the characteristic schistose and commonly banded texture of many metamorphic rocks, is due to the directed growth of the newly formed lamellar minerals. This parallel texture, which only comes into being during metamorphism, has nothing to do with the stratification of the original sedimentary rocks, and only in exceptional cases does the schistosity coincide with the sedimentary bedding. In phyllites, mica schists and gneisses, as in all schistose metamorphic rocks, the lamellar minerals lie with their tabular surfaces vertical to the direction which was subjected to maximum pressure.

Different rocks react to metamorphic conditions in very different ways. Magmatic rocks, which were formed at high temperature and pressure, are far less sensitive to temperature increases than, say, the sulphide ore minerals. Evaporites, which were formed originally at low temperatures, are particularly susceptible to changing conditions, and a small rise in temperature suffices to affect them. The more intensely a rock has been metamorphosed, the harder it is to determine what it originally was. Clues are sometimes furnished by relict minerals that have failed to react to the new conditions. Tracing the parent rocks is made even more difficult by the fact that similar metamorphic rocks can be formed from different originals. Thus in the katazone marls and basalts can both be metamorphosed to amphibolites, and gneisses of identical composition can result from deep-seated granitic rocks, volcanic rhyolites or sedimentary arkoses (felspathic sandstones). Intensive metamorphism of carbonaceous rocks produces graphite, although this mineral can also be formed hydrothermally. Emery can form from bauxite, and in many metamorphic suites epidote, serpentine, talc and the fibrous asbestos minerals are represented.

A whole series of minerals occur exclusively in metamorphic parageneses. Examples are tremolite, actinolite, glaucophane, staurolite, kyanite, sillimanite, wollastonite, scapolite, cordierite, nephrite and jadeite. Conversely, the following minerals which occur in igneous rocks are almost completely absent from metamorphic ones: sanidine, leucite and the other felspathoids, melilite and the orthorhombic pyroxenes. Even the mild metamorphism of the epizone will convert these minerals into others. Rock glass is of course also absent from metamorphic rocks. Just as there is overlap between magmatic, pegmatitic, pneumatolytic and hydrothermal deposits, there are also transitions between sedimentary or magmatic processes on the one hand and metamorphic ones on the other. For instance *diagenesis*, the consolidation of sediments into sedimentary rocks, sometimes passes into incipient metamorphism; and lavas of great geological age sometimes show what are called *anchi-metamorphic* changes in their minerals and fabric.

Although no open cavities are formed in areas of high pressure, very fine crystals may nonetheless be formed under these conditions. Many minerals find metamorphic conditions highly congenial to crystallization, and they grow into idiomorphic individuals. Crystal growth of this kind in a predominantly solid rock environment is a *blastic* process, and if the crystals assume the form proper to them they are *idioblasts*. The minerals forming crystallographically well-developed idioblasts include garnet, hornblende, magnetite, biotite, chlorite, albite, staurolite, tourmaline, rutile, kyanite, epidote and zoisite. Particularly large crystals of this kind, *porphyroblasts*, can grow if metamorphism is accompanied by supply of the appropriate chemical material. The commonest porphyroblasts are felspars.

Contact metamorphism (also called *thermal metamorphism*) requires quite different conditions. Contact metamorphic minerals and rocks are formed when a magmatic melt or migrating

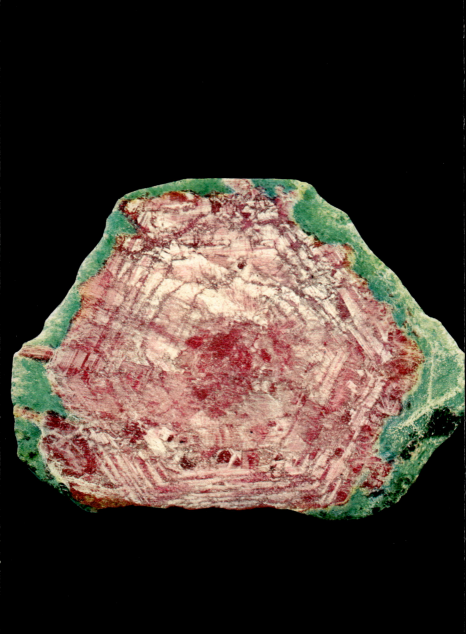

fluids (made mobile by metamorphism) come into contact with older rocks. The degree of metamorphism depends on the temperature gradient between the melt and the country rocks into which it is introduced, on the differences in chemical composition between the two, and on how long interaction between them persists. Thus few new minerals are formed when a lava flows over a sedimentary surface and quickly solidifies, baking only the uppermost layers of sediment. But it is very different when magma is intruded into sediments at depth. Under these conditions a zone of contact, a *contact aureole* is formed, where the original minerals respond to the new conditions (particularly the increase

◁ 93 RUBY blood-red corundum, with a crust of finely crystalline green zoisite. The cross-sectioned ruby crystal shows zonar banding which corresponds to its trigonal symmetry. From Longido, Tanganyika. Latin *rubeus*, red. Al_2O_3 trigonal. The red colour is due to small amounts of chromium. Zoisite after the Austrian Baron Von Zois. $Ca_2Al_3[O/OH/SiO_4/Si_2O_7]$ orthorhombic. Scale 0·7:1

94 ACTINOLITE dark green long prismatic crystals in yellow brown ▷ mica-schist. From Greiner Mountain, Tirol, Austria. Greek *aktis*, beam, and *lithos*, stone. $Ca_2(Mg, Fe)_5[(OH)_2/(Si_4O_{11})_2]$ monoclinic. Scale 1·2:1

95 STAUROLITE interpenetrant twins at right angles. From Morgan- ▷ ton, Georgia, U.S.A. Greek *stauros*, cross, and *lithos*, stone. $Al_4Fe[O/OH/SiO_4]_2$ orthorhombic. Scale 4·8:1

96 STAUROLITE interpenetrant twins at a skew angle. From Quim- ▷ per, Brittany. Scale 2·4:1

in temperature), and the result is that new minerals appear, e.g. clay minerals are converted into andalusite and chiastolite with carbonaceous inclusions (Pl. 86–7). Dense hornfelses are formed right at the junction between sediment and intrusion, where the thermal effects are most powerful. A manganese silicate also of contact metamorphic origin is rose-red rhodonite. It is often traversed by fractures infilled with black manganese oxides and hydroxides (Pl. 89). One of the rare contact minerals is sky-blue lapis lazuli (Pl. 90). Thermal metamorphism converts sandstones into quartzites, slates into spotted slates, spotted mica schists or hornfelses of one kind or another. Marls become calc-silicate rocks, and impure dolomites change to magnesium silicate rocks.

If a magmatic melt comes into contact with a sedimentary rock which contains a good deal of water, the water is driven out as superheated steam, and may carry mineral material with it which is redeposited at cooler localities. If a solidifying magma itself sends liquid or gaseous materials into the country rocks, then the *contact pneumatolytic mineral parageneses* already described are formed. But contact metamorphism is not restricted to cases where the magma supplies material to the cooler country rock. The melt may assimilate material from the adjacent sediments, and in this way different mineral suites are produced.

97 ROCK CRYSTAL (quartz) with inclusions of chlorite flakes. The ▷ upright crystal shows zoning, the included chlorite dust betraying earlier stages of growth. From Rien valley, near Göschenen, Uri, Switzerland. Rock crystal SiO_2 trigonal. Chlorite: mixed crystals of amesite and antigorite. Amesite from its occurrence in Amity, Colorado. $Mg_4Al_2[(OH)_8/Al_2Si_2O_{10}]$. Antigorite after the Antigorio valley, Piedmont. $Mg_6[(OH)_8/Si_4O_{10}]$. Both monoclinic. Scale 2:1

Rare minerals can form in contact metamorphosed coal or salt horizons, or if oxide or sulphide ores, or minerals in cap zones of sulphide lodes are subjected to these conditions. Yet other minerals arise from a reciprocal interaction between magma and country rock, or when residual melts and solutions penetrate into sediments and crystallize out there.

Metamorphic rocks in the root zones of mountains have generally been metamorphosed a number of times — they are *polymetamorphic*. This repeated recrystallization is sometimes so thorough-going that it is impossible to tell what the rock was originally. Rocks which have already been quite intensively metamorphosed can also be changed back when subjected to new and milder metamorphic conditions. This process, which results, e.g., in the conversion of biotite to chlorite minerals, is known as *regressive metamorphism*. Finally highly metamorphosed rock sequences may sink to even greater depths or come into close contact with magmas and so themselves acquire (or return to) a magmatic state. In such cases one speaks of *ultrametamorphism*, and the in part metamorphic and in part magmatic rocks formed in this way are *migmatites*. Metamorphic rock masses may be made liquid (or partially liquefied) at depth. Pegmatitic exudations may migrate and magmatic melts may permeate metamorphic rocks by seeping along their planes of schistosity. Thus metamorphic and ultrametamorphic rocks include many rock types with the most varied combinations of minerals.

Mineralogists and collectors have come from all over the world in search of the splendid crystals found in joints and fissures in the Alps. Crystals belonging to a surprisingly large number of mineral species occur with almost perfectly developed forms with sharp edges. The inaccessibility of the Alpine localities makes the quest even more attractive. These Alpine minerals were deposited from aqueous solutions which derived their mineral substance largely from the country rock adjacent to the vugs and cavities in which they are deposited. Fractures and clefts in a series of strata were formed by folding during an orogeny, and these fissures were filled with hot water rising from depth. These mineral solutions, usually containing a good deal of carbon dioxide, reacted with the rock adjacent to the cavities which were being infilled with crystals, and assimilated more soluble mineral material. When the temperature dropped, various mineral suites crystallized from the mother liquor. It follows that the manifold mineral combinations in alpine fissures vary with the adjacent country rock, and so it is often possible to infer the locality if one is given the minerals which form an assemblage there. Of course these fissure assemblages are not all of equal beauty, and with these Alpine minerals variations between one group of crystals and another is particularly common, even when they are from nearby localities.

There are still men who make it their profession to collect these assemblages, and often they alone know where to find the most rewarding places among the barren peaks. Theirs is a dangerous task, for many of the best localities lie on steep rock faces, in the paths of avalanches, in scree-filled gullies or in the permanent snow fields. Getting these delicate crystals safely down into the valley

is often as difficult as finding a lode and extracting its contents. But the mere sight of a cavity stocked with an infinite number of the finest crystals is enough to compensate for the preceding toil. Larger fissures, called *crystal cellars* or *crystal caves,* often contain many hundredweight of rock crystal and other minerals, and new finds cause as much excitement as ever. It is however often the smaller cavities that yield the rarer minerals, or ones which are particularly well crystallized. Caves long since exploited show how much mineral material has already been collected. Many of the worked-out fissures contained several hundred pounds of crystals, like the cave discovered in 1719 on the Zinggenstock in Switzerland.

The most common and characteristic mineral of alpine clefts is *rock crystal* (Pls. 97, 98, 102), which has been found in specimens up to 3 ft long and 220 lb in weight. Particularly well-grown crystals which became famous on discovery have been given individual names: for example, the 'grandfather' (293 lb), the

◁ 98 ROCK CRYSTAL (window quartz), a skeletal crystal forming 'windows'. From Porretta, Bologna, Italy. See Pl. 97. Scale 2·2:1

99 HAEMATITE rosette-like groups of thin tabular dark steely blue ▷ crystals. From Fibbia, on the Gotthard, Switzerland. Greek *haima*, blood, because of the colour of this red iron ore, and its supposed styptic power. Fe_2O_3 trigonal. Scale 3·3:1

100 FLUORSPAR (fluorite) group of pink transparent octahedra. ▷ From Grimsel, near the Gotthard, Switzerland. See Pl. 8. Scale 2·8:1

'king' (279 lb), 'Charles the Fat' (231 lb), and 'giant two points' (147 lb), are all smoky quartzes from a vein discovered in Switzerland in 1868. Smoky quartz is a smoky brown or brownish-black variety and is found high up in the Alps (above 758 ft). Jet black smoky quartz is called morion. Rock crystal in the Alps tends to be developed as prisms, but it occurs in many other forms too. Specimens are often distorted, twinned on various laws or irregularly intergrown. Spiral or curved rock crystals are composed of a large number of individuals regularly intergrown. A special type are the skeletal 'window quartzes' (Pl. 98).

The large number of minerals which go to make the assemblages of these joints and fissures include adularia (Pl. 101), albite, pericline, sphene (Pl. 103), anatase, brookite, rutile, axinite (Pl. 107) and chlorite (Pl. 103). Adularia is the name given to the transparent orthoclase of vitreous lustre which occurs in Alpine clefts. Crystals and twins of adularia are often thickly coated with chlorite, or covered with a fine layer of chlorite dust. Axinite contains boron and belongs with contact pneumatolytic calc silicate rocks. In these Alpine occurrences it usually exists as axe- or wedge-shaped crystals with knife-like edges (Pl. 107). Sphene is present as an accessory mineral in certain igneous rocks, where it has grown in position within the melts, often forming thinly tabular and minute crystals shaped somewhat like miniature envelopes. The sphene crystals of alpine clefts, however, have grown outwards from a base on which they rest, and may assume many

101 ADULARIA completely covered by chlorite flakes (see Pl. 97), ▷ with a haematite rosette at the top. From Wassen, Uri, Switzerland. Adularia after *Mons Adula*, Latin for the Gotthard. $K[(Al, Si)_4O_8]$ monoclinic. Scale 3·5:1

forms. They are of tabular or prismatic habit, and usually twinned (Pl. 103). Haematite (Pls. 99, 101) forms distinctive rosettes, which look like open flowers and are composed of intergrown thinly tabular or lamellar crystals. These haematite lamellae contain a small amount of titanium. The chlorite family is widely distributed in alpine clefts, and comprises a group of leek-green or dark-green minerals which form thin flaky coatings to rock crystal, or which may occur as inclusions in other crystals or form a rock coating on which the other minerals grow (Pls. 97, 101, 103, 107). They frequently cover the joint walls or fill vugs as loose 'chlorite sand'. The chlorite minerals penninite and clinochlore (Pl. 91) may occur as individual crystals of some appreciable size.

The bushy, fibrous or matted actinolite minerals amianthus and byssolite (Pl. 105) used to be known as 'mountain flax'. These odd-looking mineral aggregates which can often be bent consist of tiny acicular or hair-like crystals of actinolite. Joints in the calcareous Alps also contain calcite and dolomite as perfectly transparent crystals. Fluorspar is among the minerals which are rare in Alpine fissures. In some localities it occurs as beautiful octahedral crystals (Pl. 100), whereas in hydrothermal lodes it is usually developed as cubes (Pl. 53–4). Hessonite (Pl. 91), a bright red calcium aluminium garnet with resinous lustre, is sometimes found in the fissures of chlorite schists, where it forms colourful assemblages with green clinochlore or other chlorite minerals, and sometimes also with diopside. Most localities where these prized minerals of Alpine fissures occur have as country rock the magmatic and metamorphic rock complexes of the central Alps. Finds in the metamorphic calcareous Alps and Dolomites are rarer, although many minerals of other kinds do occur in these carbonate rocks. Parageneses which include sulphides with well-developed crystals, e.g. of zincblende (Pl. 104), are only formed in Alpine fissures under very specialised conditions.

Meteorites are solid bodies from outer space. Small meteorites burn out when they enter the earth's atmosphere, and may form a trail of blazing star fragments. As larger cosmic bodies pass through the atmosphere they are heated sufficiently to make a thin zone on their surfaces incandescent, but the low temperature of outer space persists in their cores. Mineral collections preserve only few of the meteorites which illuminate the sky as they race across it and crash onto the earth with loud detonations, even though they are of very special interest as tangible evidence of distant bodies. Large meteorites were once worshipped as objects from heaven. Swords and other weapons formed from them were supposed to make those who wielded them invincible.

These extra-terrestrial bodies can be divided according to their composition into *stony* and *iron* meteorites. The stony meteorites obviously represent the outer layers of the silicate shells of exploded cosmic bodies analogous with the Earth, whereas the iron meteorites probably derive partly from the cores of such bodies. Intermediate between stony and iron meteorites are the much rarer *pallasites,* consisting of cellular nickel-iron with inclusions of round olivine crystals (which have solidified from magma droplets) and other meteorite minerals (Pl. 108). These olivine crystals have a light-coloured core and darker margins. The stony meteorites almost invariably have a black crust (due to melting as they flew through the atmosphere) of the same chemical composition as the interior. The various alloys of iron-nickel which go to make iron meteorites have been given individual names and the composite structure of these iron meteorites is shown by the 'Widmannstätten figures' (named after their discoverer), formed when a polished

surface is etched by means of acid. The figures are due to the inequality of the etching action on thick and thin metal plates of different composition.

Moldavite is a volcanic glass of extra-terrestrial origin and is found in Bohemia (near the upper Moldau). It is thought to have come from a unique 'meteorite shower', composed of glassy droplets known generally as tektites (most of which are about the size of a hazel nut) possibly from distant celestial bodies. Moldavite is cut and polished as gem-stones, as are also some of the olivine crystals in stony meteorites and pallasites. Although all the chemical elements found in meteorites also occur on earth, some meteorite compounds do not exist as terrestrial minerals, e.g. yellowish-grey schreibersite (Pl. 108). Meteorites have not yet yielded any traces of organic life, nor mineral species from which the presence of life could be inferred. Their minerals show that they come from bodies with little or no water.

◁ 102 ROCK CRYSTAL group of prismatic crystals, coloured yellow at their base by a coating of limonite, while the crystal terminations are perfectly transparent. From Isère, French W. Alps. Rock crystal SiO_2 trigonal. Limonite FeOOH orthorhombic. Scale 3·2:1

103 TITANITE (sphene) yellow-brown translucent interpenetrant ▷ twin of elongated monoclinic crystals, resting on green chlorite (see Pl. 97) and themselves partly covered with fine chlorite dust. Titanite because it contains titanium, named after the Titans of Greek mythology. Sphene from Greek *sphen*, wedge, because of its crystal shape. $CaTi[O/SiO_4]$ monoclinic. Scale 3:1

Meteorites range in size from tiny particles of cosmic dust to nickel-iron masses of over 60 tons. Cosmic bodies larger than this volatilise when they strike the earth, leaving meteorite craters often of considerable size. Meteorites of any size rarely reach the earth, but the mineral content of our planet is quite considerably enriched by accumulations of cosmic dust. Because of their scarcity value and their derivation from far distant bodies of which nothing is known, meteorites are particularly treasured in museums and collections, and their occurrences have been meticulously recorded.

◁ 104 ZINCBLENDE light brown translucent sharply defined thick crystal of tabular habit, grown on sugar-like granules of white dolomite. From Lengenbach, near Imfeld, Switzerland. Zincblende ZnS cubic. Dolomite after the French mineralogist Dolomieu. $CaMg[CO_3]_2$ trigonal. Pyrite from Greek *pyr*, fire, because it gives off sparks when struck. FeS_2 cubic. Scale 7·2:1

105 AMIANTHUS (byssolite, mountain flax) long brown entangled ▷ fibres of actinolite on a base of rock crystal. From Rotlaui valley, Bernese Oberland. Greek *amiantos*, undefiled, because the fibres could not be dyed. Byssolite from Greek *byssos*, fine linen, and *lithos*, stone. Actinolite from Greek *aktis*, beam, and *lithos*. $Na_2Ca_4(Mg, Fe)_{10}[(OH)_2O_2/(Si_4O_{11})_4]$ monoclinic. Scale 1·7:1

Gem-stones are minerals with special optical properties, and are in particular of especial hardness, beauty and rarity – properties which have led to their use for ornamental or cultic purposes. The value of gem-stones fluctuates with taste and fashion. In ancient times lapis lazuli was deemed equivalent to gold; in the Middle Ages chrysolite was coveted. Turquoise came into fashion in the nineteenth century, whereas before electric light was invented garnet, which sparkles red in candle-light, was preferred. In our time minerals previously unknown have acquired the rank of gems, e.g. lilac-coloured kunzite and emerald-green hiddenite (both transparent varieties of the mineral spodumene). Rhodochrosite too has only recently come into vogue as a gem. The better-known gems are few in number, and diamond, ruby and emerald have maintained their leading positions for centuries.

A gem-stone is not necessarily very hard. Malachite is valued for its pleasent green colour and colour banding, yet it is easily scratched. Altogether all minerals which man regards as sufficiently valuable may be classed as gems. There is no longer any justification for the term 'semi-precious' stone, which used to be current.

The most striking characteristic of a gem is its colour. Few gem-stones occur only in one colour, and for most of them multiplicity of colour varieties is a characteristic property, whether the minerals concerned are perfectly transparent or opaque. Opaque stones like malachite, turquoise or lapis lazuli are so striking because of their colour and the light effects on their surface. With transparent polished stones the light is altered as it passes through them. A body colour is formed which changes with the direction of view. The nature of the illumination is an important factor with gem-stones. Ruby, emerald or pyrope have a bright sparkle in artificial light, while other gems (e.g. sapphire, amethyst or zircon) are seen to best advantage in sunlight. The colour quality of many stones can be improved by heating or radiation.

The weight of a gem-stone is given in carats, one carat being equivalent to 1/5 g or 200 milligrams. The word comes from the Arabic *kirat*, Greek *keration,* and designates the seed of the carob-tree. In antiquity these seeds, which are of very uniform size, were used as units for weighing gems.

Many gems are unprepossessing in their natural state, and only become objects of great beauty when cut and polished. Water-clear or transparent coloured gems, which are always individual crystals, are faceted. Facets are plane faces in geometrical relationship which have been artificially cut and which are not found in the natural crystal. Opaque gems usually consist of aggregates of tiny individual crystals hardly visible to the naked eye. They include kidney ore, malachite, jadeite, nephrite, rhodonite, turquoise and the numerous chalcedony minerals. Gems belonging to this group are cut for necklaces to spherical, ellipsoidal or irregular shape, with a convex surface. If they are made into other kinds of jewellery they are generally given a rounded or irregular upper surface. Stones with the lower surface plane and the upper one showing a greater or less degree of convexity are called cabochon. The effect of transparent stones depends on how the incident light is refracted in them, but stones cut into rounded pitted forms owe their optical beauty to their polished surface.

Gems of course are only varieties of various mineral species selected because of some special property or beauty. They can therefore be found in as great a diversity of mineral provinces as other varieties of minerals. However a survey of mineral occurren-

ces yielding gems shows that the localities producing any particular variety of gem are very restricted in number, and these have often been known and exploited for centuries. In the following account the gem varieties of minerals are described according to their associations and modes of occurrence. The sequence is an arbitrary one and it results among other things in the very many precious and semi-precious varieties of quartz being described in a number of different contexts just because quartz is such a common mineral produced in a wide range of different environments.

◁ 106 AMIANTHUS (actinolite) fibrous supple strands of greenish asbestos on dark green chlorite schist. From Pinzgau, Salzburg, Austria. See Pl. 105. Scale 2·4:1

107 AXINITE brown crystals, translucent at the edges, coated with ▷ fine scales of dark green chlorite (see Pl. 97). From Piz miez, Gotthard area, Switzerland. Greek *axine,* axe, because of its crystal form. $Ca_2(Fe, Mn)Al_2[OH/BO_3/Si_4O_{12}]$ triclinic. Scale 3·7:1

108 PALLASITE meteorite consisting of roundish brown olivine ▷ grains, some of which have dark marginal zones, with schreibersite and troilite; embedded in nickel iron. The slab has been sawn through and polished, and derives from the meteorite in Saskatchewan, Canada. Found in 1931. Greek *meteoros,* hovering in the air. Pallasite (named after the explorer-scientist Pallas) is a transition type between stony and nickel-iron meteorites. Schreibersite after the mineralogist Schreiber. $(Fe, Ni, Co)_3P$ tetragonal. Troilite after the Jesuit Troili. FeS hexagonal. Scale 1·3:1

DIAMOND has been brought up from great depth and was formed under high pressure. Like graphite it consists only of carbon, but has an atomic structure of its own. Its name (Greek *adamas,* invincible) underlines the fact that it is the hardest of minerals. Yet it is very brittle and easily pulverised. Uncut diamonds are found in silica-deficient rocks (kimberlites) brought to the surface by 'pipes' (clefts which extend far down into the crust). Many diamonds are found in placer deposits, where they survive as detritus from the weathering of their parent rocks. Diamond is distinguished not only by its great hardness, but also by its strong refraction and colour dispersion, which give it its fiery appearance. Uncut diamonds look unprepossessing, apart from their strong lustre. The commonest gem form of diamond is the brilliant — a faceted form in which the incident light, strongly refracted as it enters, is reflected out again by the cut faces and at the same time broken up into its constituent rainbow colours. The brilliance and colour dispersion are strongest with clear diamonds, free from

109 KUNZITE pink-violet transparent fragment of a larger crystal. ▷
From Pala, San Diego County, California. Scale 0·8:1

110 BERYL green prismatic crystal with longitudinal grooves and a partial coating of limonite. From Hopperskaja Gora, near Nerchinsk, U.S.S.R. Scale 0·7:1

111 AMAZONSTONE green crystal with thin spotted colour bands. From Custer County, Colorado, U.S.A. Scale 1:1

112 TOURMALINE green prismatic crystal with longitudinal striations and terminal faces. From Minas Gerais, Brazil. Scale 1·5:1

113 TOPAZ wine-yellow, stumpy prismatic crystal. From Topaz Rock, Schneckenstein, near Auerbach, Erzgebirge, Saxony. Scale 2·7:1

109

110

111

112

113

fractures and inclusions of particles or gas bubbles. Hence diamonds are valued according to their purity, as well as from their size and colour. Stones which appear faultless when viewed through a hand lens which magnifies ten times are said to be 'lens pure', and these are the most valuable ones. With most gems it is the strongly coloured varieties that are preferred, but among diamonds specimens which are water-clear with a tinge of blue command the greatest value. The annual world output of diamonds is about 27 million carats, but scarcely a fifth of this quantity can be used as gems. The major fraction goes to industry as 'bort' for cutting and grinding. The most important diamond fields of today are in South Africa, Brazil and Siberia.

PYROPE, a blood-red garnet (from Greek *pyropos*, fire-eyed) also originates in ultrabasic rocks and under high pressure. Pyrope used to rank among the gems of greatest value and was mined mainly in Bohemia. Today there are important fields in S. and E. Africa and the U.S.A.

A component of many basic igneous rocks is OLIVINE (Pl. 23), with its vitreous lustre, but it rarely occurs as crystals large and clear enough for cutting. As a gem it is known as *chrysolite* (Greek *chrysos*, gold; *lithos*, stone) or *peridotite* (a French term, probably from the Greek *peri*, around, and *doter*, donor). It is one of the few gems that exist in only one colour (olive-green), with but minor deviations from this tint. The most important occurrence of olivine is on the small island of Seberget in the Red Sea, off the Egyptian coast. Arizona and Burma are also sources.

LABRADORITE (Pls. 2, 18), a blue felspar with schiller structure, is also used as jewellery. Fine specimens are found on the coast of Labrador and on St. Paul's rocks in the Atlantic. It also occurs in Scandinavia and the U.S.A.

AVENTURINE FELSPARS (French *l'aventure*, chance, because glass with a red glitter, resembling this mineral, was discovered by chance) or SUNSTONES are potash or calc-soda felspars which reflect red light at their metallic surface, while the deeper layers of the crystal appear salmon red. This light effect is due to numerous very small crystal flakes of haematite which lie on certain planes within the felspar. Aventurine felspar is found in Siberia, the U.S.A. and Norway.

ROSE QUARTZ (Pl. 130), which in recent time has become a popular stone, is of pegmatitic origin, and occurs as compact masses with small amounts of manganese to give the pink colour. It is of pearly lustre, slightly opaque and often fractured, and its pale pink to reddish violet colour is often distributed as smear-like patches. Fine specimens come from Brazil, Madagascar and the U.S.A.

AMAZONSTONE (Pl. 111), so named because it is alleged to occur on the Amazon, is an opaque bluish green to verdigris-coloured MICROCLINE FELSPAR (Greek *mikros*, small; *klino*, incline, because its cleavage planes are almost perpendicular to each other). It owes its colour to copper in small quantities. Deposits occur in the Urals, Madagascar, S.W. Africa and the U.S.A.

MOONSTONE, with its silky lustre, belongs to the potash felspars. It too is of pegmatitic origin, and when it is cut as a cabochon, a milky bluish shimmer plays over its surface. The best specimens come from Ceylon.

ZIRCON (Pl. 135), mentioned as a gem by Pliny, is not uncommon, but the crystals are rarely large and worth cutting. The blue zircons are the most valuable. Yellowish red HYACINTH (Greek *hyakinthos*, a blue flower – not the hyacinth of today) is also

114

115

116

117

118

119

114 ◁ MOSS AGATE translucent chalcedony with green moss-like inclusions of hornblende asbestos. From Guernsey, Laramie County, Wyoming. Scale 0·7:1

115 BANDED JASPER chalcedony with alternating blood-red and leek-green layers. From Verckne, Uralsk, U.S.S.R. Scale 0·5:1

116 JASPER brownish red fragment of a smashed nodule. From Auggen, Württemberg. Scale 0·8:1

117 JASPER greyish-white broken nodule, red inside. From Liel, Württemberg. Scale 1·1:1

118 AGATE red banded agate with grey chalcedony inside; with country rock. From St. Egidien, Saxony. Scale 0·7:1

119 CHRYSOPRASE chalcedony coloured green by nickel silicate. From Frankenstein, Silesia. Scale 0·6:1

120 HELIOTROPE green chalcedony, spotted red. From Poona, near ▷ Bombay. Scale 1:1

121 CITRINE yellowish quartz crystals. From St. Maurice, Wallis, Switzerland. Scale 1·1:1

122 CARNELIAN dark blood-red nodule with concentrically layered structure. From Campo de Maia, Rio Grande do Sul, Brazil. Scale 0·6:1

123 AMETHYST pyramidal crystals with violet terminations. From Minas Gerais, Brazil. Scale 2·1:1

124 PETRIFIED WOOD silicified with chalcedony. The yellow and red colouration is due to limonite. From Dschebel Moka Ham, Egypt. Scale 0·7:1

120

122

121

123   124

125

126

127

128

129

125 ◁ ROCK CRYSTAL water-clear transparent quartz crystals. From Gotthard, Switzerland. Scale 0·7:1

126 CARNELIAN white and red banded nodule. From Reichenbach, near Idar-Oberstein, Germany. Scale 0·6:1

127 NEPHRITE compact aggregate of slender actinolite needles. From Wakatipu, Otaga Province, N. Zealand. Scale 0·5:1

128 SMOKY QUARTZ dark smoky-brown quartz crystals. From Val Cristallina, Tessin, Switzerland. Scale 1·1:1

129 AVENTURINE QUARTZ dense quartz aggregate with many small included flakes of haematite. From Taganai, Urals. Scale 1·2:1

130 ROSE QUARTZ coloured pink in patches, translucent at its edges. ▷ From Ambosita, Madagascar. Scale 0·5:1

131 STAR SAPPHIRE. The star of light is caused by tiny hollow canals. From Pelmadulla, Ceylon. Scale 1:1

132 TOPAZ yellow crystal with many faces. From Ouro Preto, Brazil. Scale 1·6:1

133 RUBY red corundum crystal. From Rakawana, Ceylon. Scale 5·5:1

134 DIAMOND octahedral crystal with incipient more complicated faces. From S. Africa. Scale 6:1

135 ZIRCON brown crystal in nepheline syenite (zircon syenite). From Miask, Urals. Scale 5·5:1

136 SAPPHIRE blue corundum crystal. From Ratnapura, Ceylon. Scale 7·5:1

130

131

132

133

134

135

136

prized. Faceted zircon resembles diamond in that it too has strong refraction and colour dispersion. Zircon's blue colour is due to the radioactive decomposition of thorium inclusions, and can in many cases be altered by heating the stone. Although among the most beautiful of gems, zircon will not stand up to a strong impact. The best specimens are from Indochina and Ceylon, where they are mainly won from placers.

SPODUMENE (Greek *spodios,* ash-coloured) forms grey opaque crystals in pegmatites. There are two varieties of gem quality that are transparent and have a strong vitreous lustre, namely KUNZITE (Pl. 109), named after the New York jeweller Kunz and pink, lilac or violet in colour; and HIDDENITE (after its discoverer Hidden), coloured emerald green or bright yellow-green by traces of iron and chromium. Spodumene of gem quality is found in Brazil, Madagascar and the U.S.A.

BERYL (Pl. 110) derives from pegmatitic-pneumatolytic para-geneses. In the Roman Empire colourless beryl crystals were ground as spectacle lenses to correct short-sightedness. Clear transparent beryl is called AQUAMARINE (Latin *aqua,* water; *mare,* sea). The finest aquamarines come from Brazil, Madagascar, S.W. Africa, and the Urals. Golden beryl comes from Ceylon and S.W. Africa. HELIODOR (ancient proper name: the sun's gift) is a bright yellowish green and comes from S.W. Africa. Beryls (both brightly coloured and colourless) of gem quality are also found in Australia and the Urals. The most valuable variety of beryl is green EMERALD (Pl. 35). For centuries Columbia has provided the best emeralds, although there are fine specimens from the Urals. Other localities are N. Transvaal, S. Rhodesia, India and Salzburg. In antiquity emerald was mined in Egypt. CHRYSOBERYL (Greek *chrysos,* gold) is of greenish colour and is found in Brazil, India, Ceylon and

Madagascar. A valuable variety is chrysoberyl cat's eye, and when this stone is cut and turned in the hand, a bluish silvery-white strip of light undulates across its surface. This effect is due to tiny hollow canals arranged parallel to the long axis in the crystal. Stones of this kind come from Ceylon and Brazil. The rare mineral ALEXANDRITE (named after the Czar) is a variety of chrysoberyl coloured by a small amount of chromium and found in the Urals and in Ceylon. It ranks with the most valuable gems, looks green in daylight but red in artificial light. The most precious specimens have a colour that alternates between emerald green and ruby red.

TOURMALINE can occur in many different colour varieties. Colourless crystals are called ACHROITE (Greek *achroia*, colour-lessness), red transparent ones RUBELLITE (Latin *rubellus*, reddish). Purple tourmaline is APYRITE (Greek *a*, not; *pyr*, fire, because it has less sparkle than ruby). A violet variety is SIBERITE (from its occurrence in Siberia). Green varieties of different colour intensity are particularly valued, while ordinary opaque black tourmaline or SCHORL (*schor*, impurity, because in placer deposits of tin the tourmaline occurs as an impurity in cassiterite) is rarely used as a gem-stone. Colourless, pink or green columnar crystals with dark terminal faces are called 'Moors' heads'. Tourmaline (Pls. 19, 36, 112) usually forms elongated prisms with many terminal faces and striations or grooves parallel to the direction of elongation. Most specimens of gem quality are from Brazil, Madagascar and S. and S.W. Africa.

TOPAZ is named after an island in the Red Sea, although some authorities derive the name from Sanscrit *tapas*, fire. It is of pneumatolytic origin and forms yellow, red, blue or colourless crystals, usually prismatic with many faces (Pls. 113, 132). Topaz has good cleavage and so must be cut and polished very carefully.

137

138

139

140

141

142

Crystals of dark wine-yellow colour are most popular as gems. Pink ones, formed by heating yellow specimens, are also well liked. Fine specimens are obtained from Brazil, Ceylon, Siberia and S.W. Africa.

MELANITE (Greek *melos*, black), a garnet, is found in many alkaline igneous rocks. It is sometimes used for tombstones.

Many silica minerals of gem quality are deposited in the vesicles of lavas. Often the amygdales are margined by banded CHALCEDONY, with quartz crystals protruding into the empty central space. Chalcedony is an aggregate of very fine quartz fibres (Pls. 29, 30, 31) of varying colour — white, grey, blue — which is often unevenly distributed within the stone. There are considerable deposits in Brazil, Iceland, Siberia, Syria, Transylvania and Ceylon. Banded amygdales of chalcedony with differently coloured layers of variegated structure are termed AGATES (Pls. 32, 118). Today most agates come from Brazil and Uruguay. Their natural

◁ 137 LAPIS LAZULI. From Lake Baikal, Siberia. Scale 1·2:1
138 TIGER'S EYE yellow brown aggregate of parallel fibres with silky lustre, some bands brown. From Prieska, S. Africa. Scale 0·5:1
139 FALCON'S EYE greenish grey-blue aggregate of parallel fibres with silky lustre. A fissure infilling. From Griquatown, S. Africa. Scale 0·5:1
140 ALMANDINE brownish red garnet crystal. From Zillertal, Tirol. Scale 1:1
141 ANDALUSITE green coated thick columnar crystals. From Lisenzalpe, Tirol. Scale 0·8:1
142 EPIDOTE dark green prismatic crystals. From Salzburg, Austria. Scale 0·8:1

143

144

145

146

147

148

colour is often grey, and so they are artificially coloured before being processed. ONYX (Greek *onyx*, fingernail) is a flat-banded variety, with alternating layers of black and white.

The most valuable amygdaloidal gem is AMETHYST (Pls. 1, 123). Its colour ranges from lilac to dark violet (the colour of the most prized specimens). The terminal faces are often darker than the prisms. Deposits used to be worked at Idar-Oberstein in Germany, but now the most important sources are in Brazil, Uruguay, Canada, the U.S.A., Ceylon and Madagascar. Amethyst can be coloured yellow or yellowish-brown by heating.

Quartz crystals coloured lemon-yellow or brownish by iron are called CITRINE (Pl. 121) and are found in Brazil, Spain, Madagascar, the U.S.A. and the Urals.

OBSIDIAN (after the Roman Obsidius, who was the first to bring it from Ethiopia) is sometimes fashioned into jewellery or ornaments. It is not a mineral but a glass formed by the rapid cooling of acid lavas. Green varieties are popular, and specimens come

◁ 143 AMBER with inclusions of plant remains and insects (spider and ant). From Samland coast, E. Prussia. Scale 0·6:1

144 TURQUOISE bluish aggregate of small crystals. From Nischapur, Province of Chorassan, Persia. Scale 0·6:1

145 OPAL. From White Cliffs, Australia. Scale 1·1:1

146 OPAL. From Jundah, Thomson River, Queensland, Australia. Scale 0·5:1

147 BLOOD STONE nodular botryoidal haematite. From Cumberland. Scale 0·6:1

148 SERPENTINE dense green specimen. From Wurlitz, Fichtelgebirge. Scale 0·8:1

from Mexico, Iceland, the U.S.A. and Greece. Silicified volcanic tuffs are also used as jewellery.

OPAL (Pls. 20, 145, 146) is often formed as a concomitant of volcanic processes. It occurs in a number of colours. If amorphous opal dries out it becomes fractured and loses its opalescence. In the course of geological time opal changes first to chalcedony and finally to quartz. Since the deposits in Hungary have been worked out the finest specimens have come from Mexico and Australia. Fire opal, a transparent variety occurring as nodules in rhyolite, is from Mexico.

BLOOD STONE (Pl. 147), a variety of haematite, is among the few gems that are hydrothermally formed. It has a strong steely lustre when cut and polished, and occurs as large aggregates in Cumberland, Norway, Sweden, Germany, Canada and New Zealand.

FLUORSPAR (Pls. 8, 53–4) is also mostly of hydrothermal origin. Although it occurs as well-formed and beautifully coloured crystals, its softness and good cleavage make individual crystals useless as jewels. However, massive, fine-grained aggregates such as occur in England, S.W. Africa, Australia and the U.S.A., can be fashioned into ornaments.

The hydrothermal mineral that has in recent time acquired greatest popularity as a gem is RHODOCHROSITE (Pls. 4, 55). It is a soft carbonate, but pink aggregates with colour banding can be worked into jewellery or objects of industrial use. Fine specimens are obtained from the Argentine.

JASPER is among the few gem-stone minerals of sedimentary origin. Like chalcedony it consists of silica and has a conchoidal or splintery fracture (Pls. 116, 117). Because of their hardness both jasper and flint, which is similar to it, were used to make tools and weapons. In many sedimentary rocks it forms nodular concretions which are either uniform throughout or made of alternating white and grey bands. In many localities jasper nodules are stained yellow, brown or red by weathering solutions containing iron.

CARNELIAN (Latin *carneus*, flesh-coloured; Pls. 122, 126) is a nodular or massive variety of chalcedony of a reddish or yellowish-red colour, often formed in hot dry climates as crusts, usually on sandstone. It is translucent or opaque and owes its colour to inclusions of iron oxide. There are considerable deposits of it in India, N. Africa, Brazil, Siberia and Australia. SARDONYX is a reddish-brown agate type of chalcedony, named from the ancient city of Sardis, through which the stones passed on their journey from India to Greece. HELIOTROPE (Greek *helios*, sun; *tropos*, turning, probably because the mineral was obtained in antiquitiy from Upper Egypt which lies on the Northern Tropic) or BLOOD-STONE is a sub-translucent bright green variety speckled with red from irregular inclusions of iron oxide (Pl. 120).

TURQUOISE (Pls. 82, 144) has been valued as a gem since antiquity, and bright blue specimens are regarded as of special value, although turquoise may not keep its colour. As a product of weathering, turquoise forms irregular patches which consist of firmly interlocked grains. Specimens with a network of veins infilled with yellow brown limonite and dark manganese oxides are known as turquoise matrix. The gem production comes from Persia, Afghanistan, Tibet, Australia and the Argentine.

Green MALACHITE (Pls. 9–10, 73) derives from oxidation zones of sulphide copper ore deposits in the Urals, Africa, Australia, Chile and the U.S.A. Specimens with alternating light and dark concentric bands are particularly prized and are made up of radial fibrous aggregates which have a silky lustre on surfaces broken parallel to the fibrous structure. Malachite is sometimes interbedded with AZURITE (Pls. 68, 75), forming attractive green and blue banding. Specimens are brought for polishing from S.W. Africa and Arizona. DIOPTASE (Greek *dioptano*, to look through) is not quite so suitable for ornamental purposes. It has a bright emerald-green colour but is fairly soft. CHRYSOCOLLA (Pl. 69), which is amorphous, is also rarely used as a jewel. These two substances are mined in Chile, the U.S.A. and Russia.

CHRYSOPRASE (Pls. 50, 119) forms from the weathering of nickeliferous serpentine and consists of chalcedony coloured yellowish green by tiny granules of nickel silicate embedded in it. Silesia provides particularly fine specimens.

Gem SPINEL is among the many gem-stones of metamorphic origin, and is sought as red, purple, violet and blue transparent specimens. Dark green to black spinels are named CEYLONITE (because they occur there) or PLEONASTE (Greek *pleonasmos*, superfluity). Gem spinels are found in Burma, Thailand, India, Brazil, Australia, Tasmania and Ceylon.

ANDALUSITE (from its occurrence in Andalusia) can be formed in slate through contact metamorphism. The crystals consist of thick prisms and are a turbid grey or brown and opaque. It is the rare green or red translucent andalusites from Brazil and India that are used as gems. Thick prismatic andalusite crystals with carbonaceous pigment embedded in them in the shape of a cross are called chiastolite (Pl. 86–7).

Most forms of GARNET (Latin *granum*, grain), a group which includes members of different chemical composition, are widely distributed in crystalline schists, and a few garnets are important as gems. Garnet (Pl. 140) occurs in most colours, apart from blue.

SPESSARTINE (named from its occurrence in Spessart, Germany) is a bright orange red, ANDRADITE (after the Portugese mineralogist, D'Andrada) is mostly brown, GROSSULARITE (Latin *grossularia*, gooseberry) green and transparent, DEMANTOID (a diamond-like stone) bright green, UVAROVITE (after the Russian Count Uvarov) emerald green, HESSONITE (Pl. 91) hyacinth-coloured and ALMANDINE (after the ancient town of Alabanda in Asia Minor) violet-red to wine-red. Nowadays good specimens of reddish brown hessonite and violet-red almandine come from Ceylon.

The corundum group of minerals are next in hardness to diamond, and for this reason emery, an impure corundum, is used as an abrasive. Gem corundum coloured red by traces of chromium is called RUBY (Pls. 93, 133), while SAPPHIRE (Pls. 92, 136) is the variety with a blue colour derived from iron and titanium. Sapphire may also occur in other colours. Corundum is very resistant to weathering and so occurs with other such hard and stable minerals in placer deposits. The most valuable deep red rubies come from Burma. Specimens from Ceylon are of somewhat lighter, raspberry-red colour, while rubies from Thailand have a trace of brown. The precious *star rubies* and *star sapphires* (Pl. 131) show a mysterious star of light that seems to hover over the crystal. It is caused by slender acicular inclusions. The best specimens come from Ceylon and are not faceted but cut as cabochons.

Aggregates of parallel fibres of asbestiform hornblende (Greek *asbestos*, incombustible) with interbedded thin quartz prisms are known as QUARTZ CAT'S EYE and are used for ornament. Asbestos is a name given to various mineral species which develop as almost silky long fibres. Generally asbestos is much too soft to be of any use as a gem-stone. However, some varieties of asbestiform amphibole are hardened by parallel growth with quartz, and if they are cut in cabochon and use is made of bends in the fibres, light can be concentrated in a narrow band across the cut stone which then glows with very much the appearance of a cat's eye. CROCIDOLITE (Greek *krokys*, thread; *lithos*, stone), a fibrous soda-amphibole, is also a commercial asbestos mineral and when it occurs as bluish fissure infillings with quartz it is called FALCON'S EYE. The fibres of falcon's eye are sometimes bent right over (Pl. 139). Acicular crocidolite which has weathered to a yellowish brown colour (due to the formation of limonite) is termed TIGER'S EYE (Pl. 138). Tiger's eye is thus an alteration product of falcon's eye, and in some specimens intermediate stages can be seen. Quartz cat's eye is found in Ceylon, India and Brazil, and falcon's eye and tiger's eye in S. Africa and Burma.

Flesh red to brownish red, massive, weakly translucent aggregates of RHODONITE (Pl. 89) are also used for ornamental work. Some specimens have a network of thin dark veins infilled with black manganese oxides. The mineral is found in the Urals, Sweden, Australia and the U.S.A.

Dark azure-blue lazurite or LAPIS LAZULI (Pls. 90, 137) has been a valued gem for centuries. It often forms opaque finegrained or compact aggregates with brass-yellow pyrite crystals embedded in them. Lapis lazuli lies in country rock of grey-white marble. There are extremely old workings in N.W. Afghanistan, and others in Siberia and Chile.

JADE is a collective term of specimens of JADEITE and NEPHRITE of gem-stone quality. (Jadeite from Spanish *ijada*, flank, in *piedra de ijada*, colic stone – a stone amulet against pain; nephrite from

Greek *nephros,* kidney, an amulet against kidney complaints). Jadeite is a pyroxene, nephrite (Pl. 127) an amphibole, yet they both form similar opaque and predominantly green masses of very compact minute crystal fibres which are extremely tough. Nephrite is less brittle than steel and was a valuable raw material for weapons and tools in prehistoric times. Yellowish-white and brown tints are known as well as green ones. Nephrite is highly valued in China (where it is called Yu) and serves as a raw material for ornamental work. Jadeite is mined in Burma, spinach-green nephrite in New Zealand.

AVENTURINE QUARTZ (Pl. 129) is the name given to quartz containing red spangels of mica. As aventurine quartz is turned in the hand, numerous specks glitter against the red-brown background. There is a green aventurine quartz which has inclusions of tiny flakes of the green chromium mica fuchsite (named after the mineralogist von Fuchs). Localities for aventurine quartz are the Urals, Siberia, India, China, Brazil and Madagascar.

ROCK CRYSTAL (Pls. 97, 98, 102, 125) is found in pegmatites and in the ore deposits of Alpine fissures. Today it is of little value as a gem, but has important industrial uses. Perfectly transparent specimens can be had in the Alps, Madagascar, Brazil, Alaska, Japan and the U.S.A. Blackish-brown or smoky-grey rock crystal is called SMOKY QUARTZ (Pl. 128), and perfectly black crystals are named MORION (Greek *moroeis,* dark) and are found in Brazil. Rock crystal which has inclusions of thin prisms of tourmaline or red-brown needles of rutile is favoured as a raw material for ornamental work and occurs in Switzerland, Madagascar and the U.S.A. Pistachio-green to dark green EPIDOTE (Pl. 142) occurs in fissure ores but is rarely used as a gem (Greek *epidosis,* something given as an extra). Fine specimens came formerly from Salzburg, and today from Norway, the U.S.A. and the Urals.

Among the rare extra-terrestrial substances used as gems is bottle-green or brownish MOLDAVITE (from its occurrence on the Moldau). Material of this character is found in Bohemia, Australia and the East Indies.

Finally, amber, pearl, jet and coral are also gems, although they are organic products and not minerals. AMBER (Arabic *'anbar*, ambergris) was formed from the resin of conifers growing in Lower Tertiary times on the site of the modern Baltic. The most important localities are on the coast of E. Prussia. Amorphous amber forms nodular masses which are usually honey-yellow or yellowish brown and, in their natural state, covered by a white crust. Amber is transparent, often full of fractures and with many included air and gas bubbles, conifer needles, leaves and other plant fragments and insects (Pl. 143). The creatures embedded in this way are often in an excellent state of preservation. PEARLS grow in molluscs around small foreign bodies. They consist of very small laminae of a substance called mother-of-pearl. The laminae are separated by thin skins, and this is responsible for the iridescence of the pearl. More important than the river pearl molluscs are the shell fish living in shallow sea-water within the tidal zone. Most pearls come from shell banks in the Persian Gulf, the Indian Ocean and off the coast of Australia and Central America. Pearl molluscs will deposit pearl substance round globules of mother-of-pearl placed inside them, and in this way pearls can be produced 'artificially'. Unlike crystalline gems, pearls are short lived and decompose in time as their horny substance is destroyed. The PRECIOUS CORAL of the Mediterranean is not a reef builder (as most corals are), but builds a solitary branch-like and dendritic skeleton consisting mainly of limestone, with some iron oxide to give a red colouration. JET comes from resinous wood that has been converted to a compact coal-like substance with a waxy lustre in deoxygenated mud. Deep velvet black jet which can take a high polish is used as mourning jewellery. Jet is found in the Lias of Swabia, in S. France, England, Spain and the U.S.A.

There are also synthetic gems with much the same chemical composition and properties as natural gem-stones. They are made in large numbers for industrial use, to which their properties (particularly their hardness) fits them.

LIST OF MINERALS

The following index lists the most important minerals in alphabetical order, giving their chemical formula, crystal system, density (D), hardness (H), colour (C) and occurrence (O). The names of the crystal systems are abbreviated: hex. = hexagonal; cub. = cubic; monocl. = monoclinic; rhomb. = orthorhombic; tetrag. = tetragonal; trig. = trigonal; tricl. = triclinic.

The scale of hardness is that of Mohs: 1 talc, 2 gypsum, 3 calcite, 4 fluorspar, 5 apatite, 6 felspar, 7 quartz, 8 topaz, 9 corundum, 10 diamond.

Only the principal colours are given in the case of minerals with differently coloured varieties.

The principal modes of occurrence are listed under O: alp. joints = in alpine clefts and fissures; bas. P = in basic and ultrabasic deep-seated rocks (plutons); hydrotherm. = of hydrothermal origin; magmat. = in magmatic rocks; metam. = of metamorphic origin; meteor. = in meteorites; oxidat. = in the oxidation and cementation zones of sulphide ore deposits; pegmat. = in pegmatites; pneumat. = in pneumatolytic parageneses; sal. = in salt deposits and as evaporites in regions of arid climate; sediment. = of sedimentary origin; placer = in placer deposits; sublim. = deposited from vapours in the proximity of volcanoes; volcan. = in volcanic rocks.

Alternative names are indicated by arrows, thus: ALEXANDRITE → chrysoberyl.

ACHROITE colourless
tourmaline
→ tourmaline

ACTINOLITE NaCa (Mg,
Fe)$_5$ [(OH, F)/
Si$_4$O$_{11}$]$_2$ monocl.
D: 2·9—3·1 H: 5^1/$_2$
C: green
O: metam., alp.
joints Pls. 94, 105

ADAMITE Zn$_2$ [OH/
AsO$_4$] rhomb.
D: 4·3—4·5 H: 3^1/$_2$
C: colourless, white
O: oxidat. Pl. 84

ADULARIA K[(Al,Si,)$_4$
O$_8$] monocl.
D: 2·53—2·56
H: 6 C: colourless,
white O: alp. joints
Pl. 101

AEGIRINE NaFeSi$_2$O$_6$
monocl. D: 3·5
H: 6—6^1/$_2$ C: green,
brown, black
O: magmat., volcan.,
pegmat.

AGATE SiO$_2$ trig.,
banded or patchily
coloured chalcedony
in vugs
D: 2·59—2·61
H: 6^1/$_2$ C: grey
O: volcan.
Pls. 32, 33, 118

ALABANDITE MnS cub.
D: 4·0 H: 3^1/$_2$—4
C: brownish black,
often tarnished
O: hydrotherm.

ALABASTER = fine grai-
ned compact white
variety of gypsum

ALBITE Na[AlSi$_3$O$_8$] tricl.
D: 2·62 H: 6^1/$_2$

C: colourless, white
O: pegmat., metam.,
alp. joints.

ALEXANDRITE
→ chrysoberyl

ALLANITE → orthite

ALMANDINE (iron-alu-
minium garnet)
Fe$_3$Al$_2$Si$_3$O$_{12}$ cubic
D: 4·2 H: 6^1/$_2$—7
C: red brown
O: metam., placer
Pl. 140

ALUNITE KAl$_3$[(OH)$_6$/
(SO$_4$)$_2$] trig.
D: 2·7—2·8
H: 3^1/$_2$—4 C: white
O: sediment., oxidat.

AMAZONSTONE → green
microcline, Pl. 111

AMBLYGONITE AlLi[(F,
OH)/PO$_4$] tricl.
D: 3·0—3·1 H: 6
C: white, grey,
greenish
O: pegmat., pneumat.

AMESITE Mg$_4$Al$_2$[(OH)$_8$/
Al$_2$Si$_2$O$_{10}$] monocl.
D: 2·71 H: 1^1/$_2$
C: apple-green
O: metam.

AMETHYST SiO$_2$ trig.
D: 2·65 H: 7
C: violet
O: volcan. Pls. 1, 123

AMIANTHUS = fibrous
aggregate of acti-
nolite, Pls. 105, 106

AMPHIBOLE, name for
mineral group
(hornblendes),
monocl. and rhomb.

ANALCITE Na[AlSi$_2$O$_6$]·
H$_2$O cub.
D: 2·2—2·3 H: 5^1/$_2$
C: white, grey,
yellow, red
O: volcan.,
hydrotherm.

ANATASE TiO$_2$ tetrag.
D: 3·8—3·9 H: 5^1/$_2$—6
C: brown to
blue-black
O: sediment.,
alp. joints

ANDALUSITE (chiastolite)
Al$_2$[O/SiO$_4$] rhomb.
D: 3·1—3·2 H: 7^1/$_2$
C: grey, reddish
O: metam. Pls. 86—7,
141

ANDESINE → plagioclase

ANDRADITE (calc-iron
garnet)
Ca$_3$Fe$_2$Si$_3$O$_{12}$ cub.
D 3·7 H: 6^1/$_2$
C: brown O: metam.

ANGLESITE PbSO$_4$ rhomb.
D: 6·3 H: 3
C: colourless, white,
grey O: oxidat.

ANHYDRITE CaSO$_4$
rhomb. D: 2·9—3
H: 3—4 C: colourless,
white, also coloured
O: sediment., alp.
joints, metam.,
hydrotherm.

ANKERITE → brownspar

ANNABERGITE → nickel
bloom

ANORTHITE Ca[Al$_2$Si$_2$
O$_8$] tricl.
D: 2·76 H: 6
C: colourless, white,
grey
O: magmat., volcan.

ANTHOPHYLLITE $(Mg,Fe)_7$ $[OH/Si_4O_{11}]_2$ rhomb. D: 2·9—3·2 H: $5^1/_2$ C: brownish O: metam.

ANTIGORITE fibrous serpentine $Mg_6[(OH)_8/Si_4O_{10}]$ monocl. D: 2·8—3·2 H: 2 C: leek-green to blackish green O: metam.

ANTIMONY Sb trig. D: 6·6 H: $3^1/_2$ C: tin white, often tarnished O: hydrotherm.

ANTIMONY GLANCE (antimonite, stibnite, grey antimony) Sb_2S_3 rhomb. D: 4·6—4·7 H: 2 C: lead grey O: hydrotherm Pl. 85

ANTIMONY OCHRE = mixture of antimony oxide and hydroxide. Pl. 85

APATITE $Ca_5[(OH,F)/ (PO_4)_3]$ hex. D: 3·16—3·22 H: 5 C: colourless, or coloured O: magmat. pegmat., pneumat., sediment.

APOPHYLLITE $KCa_4[F/ (Si_4O_{10})_2] \cdot 8H_2O$ tetrag. D: 2·3—2·4 H: 5 C: colourless, white, yellowish, greenish O: volcan., metam.

APYRITE = purple tourmaline

AQUAMARINE = sea blue transparent beryl

ARAGONITE $CaCO_3$ rhomb. D: 2·95 H: $3^1/_2$—4 C: colourless and coloured O: volcan., hydrotherm., oxidat., sediment. Pl. 15

ARFVEDSONITE $Na_2Fe_4Al (OH)_2Si_8O_{22}$ monocl. D: 3·4 H: $5^1/_2$—6 C: blue black O: magmat., volcan., pegmat.

ARGENTITE → silver glance

ARSENIC As trig. D: 5·5 H: 3—4 C: light lead grey, tarnishing black O: hydrotherm. Pl. 42

ARSENOPYRITE FeAsS monocl. D: 5·9—6·2 H: $5^1/_2$—6 C: bright steel grey, often tarnished O: pneumat., hydrotherm., metam.

ASBESTOS green, yellow or brown fibrous aggregates of hornblende or chrysotile

ATACAMITE $CuCl_2 \cdot 3Cu (OH)_2$ rhom. D: 3·76 H: 3—$3^1/_2$ C: dark grass green O: oxidat.

AUGITE (Na, Ca) (Mg, Fe, Al, Ti) [(Al, Si)_2 O_6] monocl. D: 3·3 H: 6 C: leek green to pitch black O: magmat., volcan., meteor. Pl. 24

AUTUNITE (lime uranite) $Ca[UO_2/PO_4]_2 \cdot 8H_2O$ tetrag. D: 3—3·2 H: 2 C: greenish yellow, yellow O: oxidat.

AVENTURINE FELSPAR (sunstone) = felspar with haematite flakes as inclusions

AVENTURINE QUARTZ = quartz with inclusions of glinting flakes. Pl. 129

AXINITE Ca_2 (Fe, Mn) $Al_2[OH/BO_3/Si_4O_{12}]$ tricl. D: 3·3 H: $6^1/_2$—7 C: brown, greenish, grey. O: pegmat., pneumat., metam. alp. joints Pl. 107

AZURITE (blue carbonate of copper) $Cu_3[(OH)_2/ (CO_3)_2]$ monocl. D: 3·7—3·9 H: $3^1/_2$—4 C: deep azure blue O: oxidat., sediment. Pls. 68, 75, 76

BARYTES (heavy spar) $BaSO_4$ rhomb. D: 4·48 H: 3—$3^1/_2$ C: white, colourless or coloured O: hydrotherm., sediment. Pls. 6, 7, 46, 53—4, 59, 77

BERYL $Be_3Al_2[Si_6O_{18}]$ hex. D: 2·63—2·80 H: $7^1/_2$—8 C: grey, colourless, greenish, sea blue: aquamarine, pink: morganite, emerald green: emerald. O: pegmat. Pls. 35, 110

BIOTITE K(Mg, Fe, Mn)$_3$ [(OH, F)$_2$/(AlSi$_3$O$_{10}$)] monocl. D: 2·8—3·2 H: 2^1/$_2$—3 C: dark brown, greenish black, black O: magmat., volcan., metam. Pl. 25

BISMUTH Bi trig. D: 9·7 H: 2^1/$_2$ C: reddish silvery white, often tarnished O: pneumat., hydrotherm.

BISMUTHINITE (bismuth glance) Bi$_2$S$_3$ rhomb. D: 6·8—7·2 H: 2—2^1/$_2$ C: tin grey O: hydrotherm.

BORACITE Mg[Cl$_6$/B$_{14}$ O$_{26}$] rhomb. D: 2·9—3 H: 7 C: colourless, grey, blue or green O: sal.

BORAX (tincal) Na$_2$B$_4$O$_7$ · 10H$_2$O monocl. D: 1·7—1·8 H: 2—2^1/$_2$ C: colourless, grey, yellowish O: sal.

BORNITE (variegated copper ore) Cu$_5$FeS$_4$ cub. D: up to 5·3 H: 3 C: reddish brass yellow, tarnishing O: pneumat., hydrotherm., sediment.

BORONATROCALCITE → ulexite

BOURNONITE PbCuSbS$_3$ rhomb. D: 5·7—5·9 H: 3 C: lead grey to iron black O: hydrotherm. Pl. 43

BRAUNITE 3Mn$_2$O$_3$ · MnSiO$_3$ tetrag. D: 4·7—4·9 H: 6—6^1/$_2$

238

C: brownish black O: metam.

BROCHANTITE Cu$_4$ [(OH)$_6$/SO$_4$] monocl. D: 3·97 H: 3^1/$_2$—4 C: dark emerald green O: oxidat.

BRONZITE (Mg, Fe) SiO$_3$ rhomb. D: 3·5 H: 5—6 C: bronze brown O: magmat.

BROOKITE TiO$_2$ rhomb. D: 4·1 H: 5^1/$_2$—6 C: yellowish red brown to black O: alp. joints

BROWNSPAR (ankerite) Ca (Fe, Mg, Mn) [CO$_3$]$_2$ trig. D: up to 3·8 H: 3^1/$_2$—4 C: yellowish brown O: hydrotherm., metam. Pls. 7, 67

BRUCITE Mg(OH)$_2$ trig. D: 2·4 H: 2^1/$_2$ C: colourless, greenish, white O: metam.

BYSSOLITE Aggregate of long fibres of actinolite. Pl. 105

BYTOWNITE → plagioclase

CALAMINE → smithsonite

CALCITE CaCO$_3$ trig. D: 2·6—2·8 H: 3 C: colourless or coloured O: volcan., hydrotherm., sediment., metam. Pls. 3, 5, 7, 13, 14, 42, 56, 67

CALOMEL (horn quicksilver) HgCl tetrag. D: 6·4—6·5 H: 1—2 C: yellow grey O: oxidat.

CANCRINITE (Na$_2$Ca)$_4$ [CO$_3$/OH/(AlSiO$_4$)$_6$] hex. D: 2·4—2·5 H: 5—6 C: colourless, yellowish O: magmat.

CARNALLITE KMgCl$_3$· 6H$_2$O rhomb. D: 1·6 H: 1—2 C: white, reddish or yellowish O: sal.

CARNOTITE (K,Na,Cu Pb)$_2$[(UO$_2$/VO$_4$)]$_2$· 3H$_2$O monocl. D: 4·5 H: 4 C: yellowish O: sediment.

CASSITERITE (tinstone) SnO$_2$ tetrag. D: 6·8—7·1 H: 6—7 C: brownish black O: pegmat., pneumat., placer

CELESTINE SrSO$_4$ rhomb. D: 3·9 H: 3—3^1/$_2$ C: colourless, white, blue, yellow O: hydrotherm., sediment. Pl. 58

CERARGYRITE (horn silver) AgCl cub. D: 5·5—5·6 H: 1^1/$_2$ C: grey, yellow brown to black O: oxidat.

CERUSSITE (white lead ore) PbCO$_3$ rhomb. D: 6·4—6·6 H: 3—3^1/$_2$ C: colourless, white, grey, brownish O: oxidat. Pl. 76

CHABAZITE $Ca[Al_2Si_4O_{12}]\cdot 6H_2O$ rhomb. D: 2·1 H: $4^1/_2$ C: white, colourless, brownish, reddish O: volcan.

CHALCANTHITE (copper vitriol) $Cu[SO_4]\cdot 5H_2O$ tricl. D: 2·2—2·3 H: $2^1/_2$ C: blue O: oxidat Pl. 72

CHALCEDONY SiO_2 trig. D: 2·59—2·61 H: 7 C: grey, white, bluish, yellowish, red O: volcan., hydrotherm., sediment. Pls. 29, 30, 31, 32, 33, 49

CHALCOCITE Cu_2S rhomb. and hex. D: 5·7—5·8 H: $2^1/_2$—3 C: dark lead grey, tarnished O: hydrotherm., oxidat.

CHALCOPYRITE (copper pyrites) $CuFeS_2$ tetrag. D: 4·1—4·3 H: $3^1/_2$—4 C: brass yellow, often tarnished O: hydrotherm., sediment. Pl. 66, 67

CHIASTOLITE = andalusite with carbonaceous inclusions in the form of a cross. Pl. 86—7

CHILE SALPETRE $NaNO_3$ trig. D: 2·2—2·3 H: $1^1/_2$ C: colourless, white O: sediment.

CHLOANTHITE (white nickel) $NiAs_{2-3}$ cub. D: 6·4—6·6 H: $5^1/_2$ C: tin white,

tarnishing grey O: hydrotherm.

CHLORITE name for mineral group consisting principally of amesite and antigorite (green minerals, usually lamellar). Pls. 97, 101, 103, 106, 107

CHLORITOID (ottrelite) $Fe_2Al_4(OH)_4[Si_4Al_4O_{20}]\cdot 2Fe(OH)_2$ monocl. D: 3·4—3·6 H: $5^1/_2$ C: dark green, black O: metam.

CHROME DIOPSIDE diopside coloured emerald green by its chromium content. Pl. 23

CHROMITE (chrome iron ore) $(Fe,Mg)Cr_2O_4$ cub. D: 4·5—4·8 H: $5^1/_2$ C: iron black O: bas. P., placer, meteor.

CHRYSOBERYL Al_2BeO_4 rhomb. D: 3·7 H: 8—$8^1/_2$ C: greenish yellow to emerald green O: pegmat., placer

CHRYSOCOLLA $CuSiO_3 \cdot nH_2O$ amorphous D: 2·22 H: 2—4 C: green O: oxidat. Pl. 69

CHRYSOLITE → olivine

CHRYSOPRASE chalcedony coloured green by nickel. Pls. 50, 119

CHRYSOTILE (fibrous serpentine) $Mg_6(OH)_6[Si_4O_{11}]\cdot H_2O$ D: 2·5—2·6 H: 3—4 C: greenish, yellowish, brownish O: metam., hydrotherm., alp. joints

CINNABAR HgS trig. D: 8·1 H: 2—$2^1/_2$ C: scarlet red O: hydrotherm.

CITRINE = rock crystal coloured yellow. Pl. 121

CLAUSTHALITE $PbSe$ cub. D: 8·3 H: $2^1/_2$ C: lead grey O: hydrotherm., pneumat.

CLINOCHLORE $(Mg,Fe,Al)_6[(OH)_8/AlSi_3O_{10}]$ monocl. D: 2·55—2·75 H: 2 C: bluish green O: metam., alp. joints Pl. 91

COBALT BLOOM → erythrite

COBALTITE $CoAsS$ cub. D: 6·2 H: $5^1/_2$ C: reddish silver white O: hydrotherm.

COLEMANITE $Ca_2B_6O_{11}\cdot 5H_2O$ monocl. D: 2·4 H: $4^1/_2$ C: colourless O: sediment.

COLUMBITE (tantalite—niobite) $(Fe,Mn)(Nb,Ta)_2O_6$ rhomb. D: 5·3—8·1 H: 6 C: brownish black O: pegmat.

COPPER Cu cub. D: 8·83 H: $2^1/_2$—3 C: copper

239

red O: hydrotherm., oxidat. Pl. 40

COPPER GLANCE
→ chalcocite

COPPER VITRIOL
→ chalcanthite

CORDIERITE (dichroite) $Mg_2Al_3[AlSi_5O_{18}]$ rhomb. D: 2·6 H: 7—7$^1/_2$ C: grey, bluish, greenish, brown O: metam.

CORUNDUM Al_2O_3 trig. D: 3·9—4·1 H: 9 C: colourless or coloured, grey O: magmat., metam. placer

COVELLITE CuS hex. D: 4·7 H: 1$^1/_2$—2 C: blue black O: oxidat. Pl. 22

CRISTOBALITE SiO_2 cub. and tetrag. D: 2·3 H: 6$^1/_2$ C: milky white O: volcan.

CROCIDOLITE $(CaNa)_{2-3}(OH)_2(Al,Fe,Mg)_5[(Si,Al)_4O_{11}]_2$ monocl. D: 3·4 H: 5$^1/_2$ C: greenish yellow O: metam.

CROCOITE $Pb[CrO_4]$ monocl. D: 5·9—6 H: 2$^1/_2$—3 C: yellow red. O: oxidat. Pl. 80

CRYOLITE Na_3AlF_6 monocl. D: 2·95 H: 2$^1/_2$—3 C: white, coloured O: pegmat.

CUPRITE Cu_2O cub. D: 5·8—6·2 H: 3$^1/_2$—4

C: red brown O: oxidat. Pl. 73

DATOLITE $Ca[OH/B/SiO_4]$ monocl. D: 2·9—3·0 H: 5—5$^1/_2$ C: colourless, white, yellowish, greenish O: pneumat., hydrotherm., alp. joints, volcan.

DEMANTOID = yellow green andradite (garnet)

DIALLAGE $CaMg(Al,Fe,Si)_2O_6$ monocl. D: 3·3 H: 6 C: greenish, grey green, brownish, black green O: bas. P., magmat., volcan.

DIAMOND C cub. D: 3·52 H: 10 C: water clear, colourless, or coloured. O: bas. P., placer Pl. 134

DIASPORE $AlOOH$ rhomb. D: 3·3—3·5 H: 6$^1/_2$—7 C: colourless and coloured O: metam., sediment.

DICHROITE → cordierite

DIOPSIDE $CaMg[Si_2O_6]$ monocl. D: 3·3 H: 6 C: colourless, grey, green, dark green O: magmat., volcan., matam., alp. joints

DIOPTASE $Cu_9[Si_6O_{18}]·6H_2O$ trig. D: 3·3 H: 5 C: emerald green O: oxidat.

DISTHENE → kyanite

DOLOMITE $CaMg[CO_3]_2$ trig. D: 2·85—2·95 H: 3$^1/_2$—4 C: colourless, white, yellowish O: hydrotherm., sediment., metam., alp. joints Pls. 7, 104

DRAVITE = brown tourmaline

DUMORTIERITE $Al_8BSi_3(OH)O_{19}$ rhomb. D: 3·3—3·4 H: 7 C: blue, brownish, red O: pegmat., pneumat.

ELEOLITE = opaque nepheline in deep-seated rocks

EMERALD = green beryl of gem quality Pl. 35

EMERY = impure corundum

ENARGITE Cu_3AsS_4 rhomb. D: 4·4 H: 3$^1/_2$ C: steel grey O: hydrotherm.

ENSTATITE $MgSiO_3$ rhomb. D: 3·1 H: 5—6 C: grey white, brownish, dark green O: magmat., pegmat.

EPIDOTE (pistacite) $Ca_2(Al,Fe)_3[OH/(SiO_4)_3]$ monocl. D: 3·3—3·5 H: 6—7 C: dark green O: metam., alp. joints Pl. 142

EPSOMITE (Epsom Salts) $MgSO_4·7H_2O$ rhomb. D: 1·68 H: 2—2$^1/_2$ C: white O: sal.

ERYTHRITE (cobalt
bloom) $Co_3[AsO_4]_2 \cdot 8H_2O$ monocl.
D: 3·07 H: 2
C: rose red O: oxidat.

FAHLERZ → tetrahedrite

FLINT = jasper

FLUORSPAR CaF_2 cub.
D: 3·1—3·2 H: 4
C: water clear,
various colours
O: pneumat., hydro-
therm., alp. joints
Pls. 8, 53-4, 59, 100

FRANKLINITE (Zn,Mn)
$(Fe,Mn)_2O_4$ cub.
D: 5·2 H: $6^1/_2$
C: iron black
O: metam.

FREIBERGITE = fahlerz
rich in silver

FUCHSITE (chromium-
bearing variety of
muscovite) $K(Al,Cr)_2$
$(OH,F)_2[AlSi_3O_{10}]$
monocl. D: 2·85 H: 2
C: emerald green
O: metam.

GAHNITE $ZnAl_2O_4$ cub,
D: 4·3 H: 8 C: dark
green O: metam.,
placer

GALENA PbS cub.
D: 7·2—7·6 H: $2^1/_2$
C: lead grey
O: hydrotherm., sedi-
ment. Pls. 3, 39, 43

GALMEI = mixture of
various zinc oxi-
dation minerals

GARNET mineral group
→ almandine
(Pl. 140), andradite,
demantoid, grossu-
larite, hessonite,
melanite, pyrope,
spessartine uvarovite.

GARNIERITE $(Ni,Mg)_6$
$[(OH)_8/Si_4O_{10}]$
monocl. D: 2·2—2·7
H: 2—4 C: emerald
green O: oxidat.,
sediment. Pl. 88

GEHLENITE → melilite

GERMANITE Cu_6FeGeS_8
cub. D: 4·6 H: 3
C: dark violet pink
O: hydrotherm.

GEYSERITE = siliceous
sinter

GIBBSITE → hydrargillite

GLAUBERITE Na_2Ca
$(SO_4)_2$ monocl.
D: 2·7—2·8 H: $2^1/_2$—3
C: colourless or
coloured O: sal.

GLAUBER SALT (mira-
bilite) $Na_2SO_4 \cdot 10H_2O$
monocl. D: 1·49 H: 1
C: colourless, white
O: sal.

GLAUCOPHANE Na_2Mg_3
$Al_2(OH)_2Si_8O_{22}$
monocl. D: 3·1
H: $5^1/_2$ C: blue grey
O: metam.

GOETHITE FeOOH
rhomb. D: 3·8—4·3
H: 5—$5^1/_2$ C: yellow
brown, red to
brownish black
O: volcan., oxidat.,
sediment.

GOLD Au cub. D: 19·23
H: $2^1/_2$—3 C: gold
yellow O: hydro-
therm., placer
Pls. 11, 41, 45

GOSLARITE (white
vitriol) $ZnSO_4 \cdot 7H_2O$
rhomb. D: 2·0
H: 2—$2^1/_2$ C: whitish,
yellowish, grey
O: oxidat.

GRAPHITE C hex.
D: 2·2 H: 1 C: dark
steely grey O: metam.,
pegmat.

GREENOCKITE CdS hex.
D: 4·9—5·0 H: 3—$3^1/_2$
C: yellow O: oxidat.

GROSSULARITE (calcium
aluminium garnet)
$Ca_3Al_2[SiO_4]_3$ cub.
D: 3·5 H: $6^1/_2$
C: greenish, yellow
red, reddish O: metam.

GRUNERITE $Fe_7(OH)Si_8$
O_{22} monocl.
D: 3·6 H: $5^1/_2$
C: brownish
O: metam.

GYPSUM $CaSO_4 \cdot 2H_2O$
monocl. D: 2·3—2·4
H: 2 C: colourless,
white, also coloured
O: sal., sediment.

HAEMATITE (specular
iron, kidney ore)
Fe_2O_3 trig. D: 5·2—5·3
H: 6 C: steely grey
to blue black, black
red, red brown
O: pneumat., hydro-
therm., sediment.,
metam. Pls. 6, 99,
101, 147

HALITE (rock salt) NaCl
cub. D: 2·1 H: 2
C: colourless, white,
blue, reddish,
yellowish, grey
O: sediment., sal.,
sublim. Pls. 16, 61

HALLOYSITE $Al_4[OH]_8/$
$Si_4O_{10}]·4H_2O$ monocl.
D: 2·0—2·2 H: $1^1/_2$
C: white, blue,
greenish, yellow-
brown O: hydro-
therm., oxidat.,
sediment.

HARMOTOME (BaK_2)
$[Al_2Si_6O_{16}]·6H_2O$
monocl. D: 2·44—2·5
H: $4^1/_2$ C: white
O: hydrotherm.,
volcan.

HAUSMANNITE Mn_3O_4
tetrag. D: 4·7 H: $5^1/_2$
C: brown, iron black
O: metam.

HAUYNE $(Na,Ca)_{8-4}$
$[(SO_4)_{2-1}/(AlSiO_4)_6]$
cub. D: 2·5 H: 5—6
C: blue, grey, white
O: volcan. Pl. 26

HEDENBERGITE $CaFeSi_2$
O_6 monocl. D: 3·55
H: 6 C: blackish
green O: megmat.,
metam.

HELIODOR = yellowish
green beryl

HELIOTROPE, bloodstone
(jasper) = green
chalcedony with red
spots. Pl. 120

HEMIMORPHITE (zinc
silicate ore)
$Zn_4[(OH)_2/Si_2O_7]·$

H_2O rhomb.
D: 3·3—3·5 H: 5
C: colourless, white,
yellowish brown,
greenish O: oxidat.

HERCYNITE $FeOAl_2O_3$
cub. D: 3·5 H: 8
C: black O: bas. P.,
magmat., metam.

HESSONITE = hyacinth
red grossularite. Pl. 91

HEULANDITE $(Ca[Al_2Si_7$
$O_{18}]·6H_2O$ monocl.
D: 2·2 H: $3^1/_2$—4
C: colourless, whitish
grey, brown, red
O: volcan. Pl. 27

HIDDENITE = yellow or
green spodumene

HORNBLENDE $(Na, Ca)_2$
$(Mg,Fe,Mn,Al,Ti)_5$
$[(OH,F)/(Si,Al,Ti)_4$
$O_{11}]_2$ monocl.
D: 2·9—3·4 H: 5—6
C: green to black
O: magmat., volcan.,
metam.

HORNSTONE = grey to
black cryptocrystalline
silica

HYACINTH = red zircon

HYALITE $SiO_2·nH_2O$
amorphous D: 2·1
H: $5^1/_2$—6 C: colour-
less O: volcan.

HYDRARGILLITE (gibbsite)
$Al(OH)_3$ monocl.
D: 2·3—2·4 H: $2^1/_2$
C: colourless, white
O: sediment.

HYDROMAGNESITE Mg_5
$[OH/(CO_3)_2]_2·4H_2O$

monocl. D: 2·2
H: $3^1/_2$ C: white
O: oxidat.

HYDROPHANE = variety
of opal

HYDROZINCITE Zn_5
$[(OH)_3/(CO_3)_2]$
monocl. D: up to 3·8
H: 2—$2^1/_2$ C: white
O: oxidat.

HYPERSTHENE (Fe,Mg)
SiO_3 rhomb. D: 3·5
H: $5^1/_2$ C: blackish
brown, blackish green,
pitch black
O: magmat., volcan.

ICE H_2O hex. D: 0·9175
H: $1^1/_2$ C: colourless
O: sediment.

ILMENITE $FeO·TiO_2$ trig.
D: 4·5—5 H: 5—6
C: blackish brown.
O: magmat., volcan.,
pegmat., placer, alp.
joints

ILVAITE (lievrite) $CaFe_3$
$[OH/(SiO_4)_2]$ rhomb.
D: 4·1 H: $5^1/_2$—6
C: brownish black
O: pneumat., metam.

INDICOLITE = indigo-
blue tourmaline

IRON Fe cub. D: 7·88
H: 4—5 C: iron grey
O: by reduction of
basalt with carbon-
aceous country rocks,
meteor.

JADE collective term for
jadeite and nephrite

242

JADEITE NaAlSi$_2$O$_6$
monocl. D: 3·2—3·3
H: 6—6^1/$_2$ C: green,
greenish white
O: metam.

JAMESONITE Pb$_4$FeSb$_6$S$_{14}$
monocl. D: 5·63 H: 2
C: lead grey
O: hydrotherm.

JAROSITE K$_2$SO$_4$·3Fe$_2$
SO$_4$·6H$_2$O trig.
D: 3·1—3·3 H: 3—4
C: yellow, ochre
brown O: oxidat.,
sediment.

JASPER = cryptocrystal-
line quartz with
splintery fracture.
Pls. 116, 117

JORDANITE Pb$_4$As$_2$S$_7$
monocl. D: 6·4 H: 3
C: dark lead grey
O: hydrotherm.

KAINITE KMg[Cl/SO$_4$]·
3H$_2$O monocl.
D: 2·1 H: 3
C: white, yellowish,
grey O: sal.

KAOLINITE Al$_4$[(OH)$_8$/
Si$_4$O$_{10}$] monocl.
D: 2·6 H: 2
C: white, yellowish,
greenish O: hydro-
therm., sediment.

KERNITE (rasorite)
Na$_2$B$_4$O$_7$·4H$_2$O
monocl. D: 1·95
H: 2^1/$_2$ C: white
O: metam.

KIESERITE MgSO$_4$·H$_2$O
monocl. D: 2·57
H: 3^1/$_2$ C: colourless,
white, yellowish
O: sal.

KUNZITE = pink
spodumene Pl. 109

KYANITE Al$_2$[O/SiO$_4$]
tricl. D: 3·6—3·7
H: 4 or 7 C: blue,
white, grey, pink
O: metam.

LABRADORITE → plagio-
clase. Pls 2, 18

LAPIS LAZULI (Na,Ca)$_8$
[(SO$_4$,S,Cl)$_2$/
(AlSiO$_4$)$_6$] cub.
D: 2·4—2·9 H: 5—6
C: azure blue
O: metam. Pls. 90, 137

LAUMONTITE Ca[Al$_2$Si$_2$
O$_6$]$_2$·4H$_2$O monocl.
D: 2·25—2·35
H: 3—3^1/$_2$ C: white
O: volcan., hydro-
therm.

LAZULITE (Mg,Fe)Al$_2$
[OH/PO$_4$]$_2$ monocl.
D: 3·0 H: 5—6
C: blue white, blue
O: metam.

LEADHILLITE Pb$_4$[(OH)$_2$/
SO$_4$(CO$_3$)$_2$] monocl.
D: 6·5 H: 2^1/$_2$
C: white O: oxidat.

LEPIDOLITE K$_2$(Li,Al)$_3$
(F,OH,O)$_2$[AlSi$_3$O$_{10}$]
monocl. D: 2·8—2·9
H: 2^1/$_2$ C: pink,
white, greenish, grey
O: pegmat., pneumat.

LEUCITE KAlSi$_2$O$_6$ cub.
D: 2·5 H: 5^1/$_2$—6
C: white, grey
O: volcan.

LIMONITE hydrous ferric
oxide, brown
haematite

LINARITE PbCu [(OH)$_2$/
SO$_4$] monocl.
D: 5·3—5·5 H: 2^1/$_2$
C: azure blue
O: oxidat.

LINNAEITE Co$_3$S$_4$ cub.
D: 4·5—4·8
H: 4^1/$_2$—5^1/$_2$ C: steely
grey O: hydrotherm.

LUDWIGITE Mg$_3$ (Fe,Mg)
Fe$_2$[BO$_3$/O$_2$]$_2$ rhomb.
D: 4 H: 5 C: black
O: metam.

MAGNESITE MgCO$_3$ trig.
D: 3·0 H: 4—4^1/$_2$
C: colourless, white or
coloured O: metam.

MAGNETITE Fe$_3$O$_4$ cub.
D: 5·2 H: 5^1/$_2$
C: iron black O: bas.
P., magmat., volcan.,
metam.

MALACHITE Cu$_2$[(OH)$_2$/
CO$_3$] monocl. D: 4
H: 4 C: dark green
to emerald green
O: oxidat., sediment.
Pls. 9—10, 66, 68, 73,
75

MANGANESE SPAR
→ rhodochrosite

MANGANITE MnOOH
rhomb. D: 4·3—4·4
H: 4 C: brownish
black O: volcan.,
oxidat.

MARCASITE FeS$_2$ rhomb.
D: 4·8—4·9 H: 6—6^1/$_2$
C: greenish brass
yellow O: hydro-
therm., sediment.
Pl. 60

MEERSCHAUM (sepiolite) approx. $Mg_4(OH)_2$ $[Si_2O_5]_3 \cdot 3H_2O$, crystalline and amorphous D: (without pores) 2 H: $2—2^{1/2}$ C: white, grey, greenish, yellowish O: metam.

MELANITE $(Ca,Ti)_3$ $(Fe,Ti)_2Si_3O_{12}$ cub. D: $3\cdot8—4\cdot1$ H:$6^{1/2}$ C: velvet black O: volcan.

MELANTERITE (iron vitriol) $Fe(SO_4) \cdot 7H_2O$ monocl. D:$1\cdot9$ H:2 C: pale green yellowish, O:oxidat. Pl. 72

MELILITE (gehlenite-åkermanite) $(Na,Ca)_2$ $(Mg,Al)_2(Si,Al)_2O_7$ tetrag. D: $2\cdot9—3\cdot0$ H: $5—5^{1/2}$ C: colourless, yellow, brown, grey O: volcan.

MERCURY Hg trig. D: $13\cdot6$ H: liquid C: tin white, often with a grey skin O: oxidat.

MICROCLINE $K[AlSi_3O_8]$ tricl. D: $2\cdot54 — 2\cdot57$ H: 6 C: white, yellowish, reddish O: magmat., pegmat.

MILLERITE (hair pyrites) NiS trig. D: $5\cdot3$ H: $3^{1/2}$ C: brass yellow O: hydrotherm., oxidat.

MIMETITE (green lead ore) $Pb_5Cl(AsO_4)_3$ hex. D: $7\cdot1$ H: $3^{1/2}—4$ C: honey

yellow, green, grey O: oxidat.

MIRABILITE → glauber salt

MOLYBDENITE MoS_2 hex. D: $4\cdot7—4\cdot8$ H: $1—1^{1/2}$ C: lead grey O: pneumat., hydrotherm. Pl. 38

MONAZITE (Ce,La,Th) $[PO_4]$ monocl. D: $4\cdot8—5\cdot5$ H: $5—5^{1/2}$ C: yellow brown, dark brown O: magmat., pegmat., alp. joints, placer

MONTICELLITE CaMg $[SiO_4]$ rhomb. D: $3\cdot2$ H: 5 C: colourless, white, yellowish O: volcan., metam.

MONTMORILLONITE Al_2 $[(OH)_2/Si_4O_{10}] \cdot nH_2O$ rhomb. D: $1\cdot7—2\cdot7$ H: 1 C: white, coloured O: pegmat., sediment., volcan.

MOONSTONE = adularia of gem quality

MORION = pitch black smoky quartz

MUSCOVITE KAl_2 $[(OH,F)_2/AlSi_3O_{10}]$ monocl. D: $2\cdot78—2\cdot88$ H: $2—2^{1/2}$ C: colourless, yellow, red, green O: magmat., pegmat., hydrotherm., metam. Pls. 21, 34

NAGYAGITE (black tellurium) Au(Pb,Sb, $Fe)_8(Te,S)_{11}$ monocl.

D: $7\cdot4$ H: $1—1^{1/2}$ C: dark lead grey O: hydrotherm.

NATROLITE $Na_2[Al_2Si_3$ $O_{10}] \cdot 2H_2O$ rhomb. D: $2\cdot2—2\cdot4$ H: $5—5^{1/2}$ C: yellowish, white O: volcan. Pl. 28

NEPHELINE $NaAlSiO_4$ hex. D: $2\cdot6—2\cdot65$ H: $5^{1/2}—6$ C: colourless, white O: magmat., volcan.

NEPHRITE $Ca_2(Mg,Fe)_5$ $(OH,F)_2Si_8O_{22}$ monocl. D: 3 H: 6 C: leek green, greenish grey O: metam. Pl. 127

NICCOLITE NiAs hex. D: $7\cdot7—7\cdot8$ H: $5^{1/2}$ C: copper red O: hydrotherm

NICKEL BLOOM (annabergite $Ni_3[AsO_4]_2 \cdot$ $8H_2O$ monocl. D: $3—3\cdot1$ H: 2 C: grass green O: oxidat.

NIOBITE → columbite

NITRE (saltpetre) KNO_3 rhomb. D: $1\cdot9—2\cdot1$ H: 2 C: colourless, white, grey O: sediment., sal.

NONTRONITE $Fe_2(OH)_2$ $[Si_4O_{10}] \cdot nH_2O$ monocl. D: $2\cdot3$ H: 1 C: yellow, whitish, yellowish green O: hydrotherm., oxidat.

NOSEAN $Na_8[SO_4/$ $(AlSiO_4)_6]$ cub.

D: 2·4 H: 5—6
C: white, grey, brown
O: magmat., volcan.

OLIGOCLASE
→ plagioclase

OLIVENITE $Cu_2[OH/AsO_4]$ rhomb. D: 4·3
H: 3 C: dark olive
green to greenish
black O: oxidat.

OLIVINE (peridot,
chrysolite) $(Mg,Fe)_2$
$[SiO_4]$ rhomb.
D: 3·2—4·2 H: $6^1/_2$—7
C: olive green
O: magmat., volcan.,
metam., meteor.
Pls. 23, 108

ONYX = banded white
and black chalcedony

OPAL $SiO_2 \cdot nH_2O$
amorphous D: 2·1—2·2
H: $5^1/_2$—$6^1/_2$
C: white or coloured
O: volcan., hydro-
therm., sediment.
Pls. 20, 145, 146

ORPIMENT As_2S_3 monocl.
D: 3·48 H: $1^1/_2$—2
C: lemon yellow
O: hydrotherm.,
oxidat. Pl. 56

ORTHITE (allanite)
$(Ca,Ce,La,Na)_2$
$(Al,Fe,Mg,Mn)_3$
$[OH/(SiO_4)_3]$ monocl.
D: 3—4·2 H: $5^1/_2$
C: brownish black
O: pegmat., volcan.,
metam.

ORTHOCLASE $K[AlSi_3O_8]$
monocl. D: 2·53—2·56
H: 6 C: white, red,

yellowish O: magmat.,
volcan., pegmat.,
metam. Pl. 19

OTTRELITE → chloritoid

PANDERMITE $Ca_2B_5O_9$
$OH \cdot 3H_2O$ tricl.
D: 2·26—2·3 H: 3
C: white O: sal.

PARAGONITE $(Na,K)Al_2$
$(OH)_2[AlSi_3O_{10}]$
monocl. D: 2·8—2·9
H: 2 C: white,
greenish O: metam.

PECTOLITE $(Ca,Na,Mn)_3$
$[Si(O,OH)_3]_3$ monocl.
D: 2·8 H: 5
C: white O: volcan.,
hydrotherm.

PENNINITE $Mg_5(Mg,Al)$
$[(OH)_8/(Al,Si)Si_3O_{10}]$
monocl. D: 2·5—2·6
H: $2^1/_2$ C: green,
bluish green
O: volcan., metam.,
alp. joints

PENTLANDITE $(Fe,Ni)_9S_8$
cub. D: 4·6—5
H: $3^1/_2$—4
C: brownish steely
grey O: bas. P.,
magmat.

PERICLASE MgO cub.
D: 3·7—3·9 H: 6
C: colourless, white
O: metam.

PERICLINE = albite in
pegmatitic vugs and
in alpine joints

PERIDOT → olivine

PEROVSKITE $CaTiO_3$
monocl. D: 4·0

H: $5^1/_2$ C: orange,
brown to black
O: volcan., magmat.,
metam.

PETALITE $LiAlSi_4O_{10}$
monocl. D: 2·4
H: $6^1/_2$ C: colourless,
reddish O: pegmat.

PHILLIPSITE $(Ca,K_2,Na_2)_2$
$[Al_4Si_{11}O_{30}] \cdot 10H_2O$
monocl. D: 2·2
H: $4^1/_2$ C: colourless,
white, grey, yellowish
O: volcan.

PHLOGOPITE KMg_3
$(F,OH)_2[AlSi_3O_{10}]$
monocl. D: 2·75—2·97
H: 2 C: brownish
red O: pneumat.,
pegmat.

PHOSGENITE (horn lead)
$Pb_2[Cl_2/CO_3]$ tetrag.
D: 6—6·3 H: $2^1/_2$—3
C: white, grey,
yellowish O: oxidat.

PHOSPHOCHALCITE
$Cu_5[(OH)_2/PO_4]_2$
monocl. D: 4·34 H:
$4^1/_2$ C: dark green O:
oxidat. Pl. 74

PICOTITE (chrome spinel)
$(Mg,Fe)(Al,Cr,Fe)_2O_4$
cub. D: 4·1 H: 8
C: black O: bas. P.,
magmat., volcan.

PIEDMONTITE = red
manganiferous epidote

PISTACITE → epidote

PITCHBLENDE (uraninite)
UO_2 cub. D: 9—10·6
H: 4—6 C: greenish,
pitch black
O: pegmat., hydro-
therm.

PLAGIOCLASE mixed-crystal series between albite Na[AlSi$_3$O$_8$] and anorthite Ca[Al$_2$Si$_2$O$_8$]: albite, oligoclase, andesine, labradorite, bytownite, anorthite, tricl. D: 2·61–2·77 H: 6–6^1/$_2$ C: colourless, white, grey, greenish, yellowish O: magmat., volcan., pegmat., metam.

PLATINUM Pt cub. D: 21·5 H: 4–4^1/$_2$ C: steely grey to silver white O: bas P., placer

PLEONASTE (Ceylonite) (Mg,Fe) (Al,Fe)$_2$O$_4$ cub. D: 3·5 H: 8 C: brownish black, blackish green O: metam.

POLYBASITE (Ag,Cu)$_{16}$ Sb$_2$S$_{11}$ monocl. D: 6·2 H: 1^1/$_2$–2 C: iron black O: hydrotherm.

POLYHALITE K$_2$Ca$_2$Mg (SO$_4$)$_4$·2H$_2$O tricl. D: 2·77 H: 3–3^1/$_2$ C: reddish, brick red O: sal.

PREHNITE Ca$_2$Al$_2$[(OH)$_2$/ Si$_3$O$_{10}$] rhomb. D: 2·8–3 H: 6–6^1/$_2$ C: colourless, white, greenish O: magmat., metam., alp. joints

PROCHLORITE (Mg,Fe)$_4$ Al$_2$[(OH)$_8$/AlSi$_3$O$_{10}$] monocl. D: 2·75–2·9 H: 2 C: saturated green O: metam., alp. joints

PROUSTITE (light red silver ore) Ag$_3$AsS$_3$ trig. D: 5·57 H: 2^1/$_2$ C: scarlet red O: hydrotherm. Pl. 47

PSILOMELANE mainly MnO$_2$ rhomb. D: approx. 5 H: up to 6 C: black O: oxidat., sediment. Pls. 12, 55

PYRARGYRITE (dark red silver ore) Ag$_3$SbS$_3$ trig. D: 5·85 H: 2^1/$_2$–3 C: dark red O: hydrotherm.

PYRITE (iron pyrites) FeS$_2$ cub. D: 5·0–5·2 H: 6–6^1/$_2$ C: brass yellow O: hydrotherm., sediment., metam. Pl. 90

PYROLUSITE MnO$_2$ D: about 5 H: up to 6 C: grey black O: oxidat., sediment. Pl. 63

PYROMORPHITE (green lead ore) Pb$_5$[Cl/ (PO$_4$)$_3$] hex. D: 6·7–7 H: 3^1/$_2$–4 C: brownish, orange O: oxidat. Pl. 78–9

PYROPE (magnesium aluminium garnet) Mg$_3$Al$_2$Si$_4$O$_{12}$ cub. D: 3·5 H: 6^1/$_2$–7 C: blood red O: bas. P., magmat., placer

PYROPHYLLITE Al$_2$(OH)$_2$ [Si$_4$O$_{10}$] rhomb. D: 2·8 H: 1^1/$_2$ C: white, yellow, green O: hydrotherm., metam.

PYROXENES mineral group including augites, monocl., rhomb.

PYRRHOTITE Fe$_{11}$S$_{12}$ hex. D: 4·6 H: 4 C: brownish, bronze coloured O: bas. P., pneumat., hydrotherm., metam.

QUARTZ SiO$_2$ trig. D: 2·65 H: 7 C: colourless (rock crystal) and coloured O: magmat., volcan., pegmat., pneumat., hydrotherm., sediment., placer, metam., alp. joints Pls. 11, 37, 39, 41, 44, 45, 52, 66, 85

QUICKSILVER → mercury

RASORITE → kernite

REALGAR AsS monocl. D: 3·5–3·6 H: 1^1/$_2$–2 C: red O: hydrotherm. Pls. 42, 56

RHODOCHROSITE MnCO$_3$ trig. D: 3·3–3·6 H: 4 C: red, brownish, raspberry red O: hydrotherm., oxidat. Pls. 4, 55, 57

RHODONITE (Mn,Fe,Ca) [SiO$_3$] D: 3·4–3·68 H: 6 C: pink to brownish red O: sediment., metam. Pl. 89

RIEBECKITE Na$_2$Fe$_4$ (OH)$_2$Si$_8$O$_{22}$ monocl.

D: 3·4 H: $5^1/_2$
C: greenish, greenish black O: magmat., volcan.

ROCK CRYSTAL SiO_2 trig. D: 2·6 H: 7
C: colourless, transparent O: pegmat., hydrotherm., alp. joints Pls. 97, 98, 102, 125

ROSE QUARTZ pink quartz (pegmat.)
Pl. 130

RUBELLITE → red or pink tourmaline

RUBY = red corundum of gem quality.
Pls. 93, 133

RUTILE TiO_2 tetrag.
D: 4·2—4·3 H: $6—6^1/_2$
C: brownish red, black O: pegmat., metam., alp. joints, placer

SAGENITE lattice-like needles of tiny rutile crystals

SAL AMMONIAC NH_4Cl cub. D: 1·53 H: 1—2
C: colourless, also coloured O: sublim.

SALTPETRE → nitre

SANIDINE $K[AlSi_3O_8]$ monocl. D: 2·57 H: 6
C: colourless and glassy, white O: volcan.

SAPPHIRE = colourless or blue corundum of gem quality.
Pls. 92, 136

SARD = brown chalcedony

SARDONYX = banded white and red or brown chalcedony

SASSOLINE (native boric acid) $B(OH)_3$ tricl.
D: 1·45 H: 1
C: white O: volcan., sublim.

SCAPOLITE mixed-crystal series of marialite $Na_4Al_3Si_9O_{24}·Cl$ and meionite $Ca_4Al_6Si_6O_{24}CO_3$, both tetrag.
D: 2·54—2·77 H: 5—6
C: white, grey, greenish, red O: pneumat., metam.

SCHEELITE $Ca[WO_4]$ tetrag. D: 5·9—6·1
H: $4^1/_2$—5 C: grey white, pale yellow O: pneumat., hydrotherm. Pl. 37

SCHORL = black tourmaline

SCHREIBERSITE $(Fe,Ni,Co)_3P$ tetrag. D: 7·1
H: 6 C: white, tarnishing yellow O: meteor. Pl. 108

SENARMONTITE Sb_2O_3 cub. D: 5·2—5·3
H: 2 C: white to grey O: oxidat.

SEPIOLITE → meerschaum

SERICITE small thin flakes of muscovite

SERPENTINE fine grained crystalline aggregates of chrysotile Mg_6 $(OH)_6[Si_4O_{11}]·H_2O$ and antigorite Mg_6 $[(OH)_8/Si_4O_{10}]$, both monocl. D: 2·5—2·6
H: 3—4 C: green O: metam. Pl. 148

SIDERITE (spathose iron) $FeCO_3$ trig.
D 3·7—3·9 H: 4—$4^1/_2$
C: pea yellow, brown to blue black
O: pegmat., hydrotherm., sediment.
Pl. 43

SILLIMANITE Al_2SiO_5 rhomb. D: 3·2
H: 6—7 C: white, grey, yellowish, brownish O: metam.

SILVER Au cub. D: 10·5
H: $2^1/_2$—3 C: silver white, often tarnished O: hydrotherm., oxidat. Pl. 46

SILVER GLANCE (argentite) Ag_2S monocl. and cub. D: 7·3
H: 2—$2^1/_2$ C: dark lead grey O: hydrotherm., oxidat.

SMALTITE (tin white cobalt) $CoAs_{2-3}$ cub.
D: 6·4—6·6 H: $5^1/_2$
C: tin white, tarnishing grey O: hydrotherm.

SMITHSONITE $ZnCO_3$ trig. D: 4·3—4·5
H: 5 C: grey, coloured O: oxidat.
Pl. 83

SODA $Na_2CO_3·10H_2O$ monocl. D: 1·42—1·47
H: 1—$1^1/_2$ C: colourless, yellowish white, grey O: sal.

SODA NITRE → Chile saltpetre

SODALITE $Na_8[Cl_2/(AlSiO_4)_6]$ cub. D: 2·3 H: 5 C: white, grey O: magmat., volcan.

SPERRYLITE $PtAs_2$ cub. D: 10·6 H: 6—7 C: tin white O: bas. P., magmat., pegmat.

SPESSARTINE (manganese aluminium garnet) $Mn_3Al_2Si_3O_{12}$ cub. D: 4·2 H: $6^1/_2$ C: yellow, red brown O: magmat., metam.

SPHALERITE (zincblende, blende, black jack) ZnS cub. D: 3·9—4·2 H: $3^1/_2$—4 C: yellow, reddish, brown, black O: hydrotherm., sediment. Pls. 3, 44, 48, 51, 104

SPHENE (titanite) $CaTi[O/SiO_4]$ monocl. D: 3·4—3·6 H: 5—$5^1/_2$ C: yellowish, brown, greenish, blackish brown O: metam., alp. joints Pl. 103

SPINEL $MgAl_2O_4$ cub. D: 3·5 H: 8 C: colourless or coloured O: metam., placer

SPODUMENE $LiAlSi_2O_6$ monocl. D: 3·1—3·2 H: $6^1/_2$—7 C: grey, pink, yellowish green, colourless O: pegmat.

STANNINE Cu_2FeSnS_4 tetrag. D: 4·3—4·5

H: 4 C: olive steely grey O: pneumat., hydrotherm.

STAUROLITE $Al_4Fe[O/OH/SiO_4]_2$ rhomb. D: 3·7—3·8 H: 7—$7^1/_2$ C: blackish brown O: metam. Pls. 95, 96

STEATITE → talc

STEPHANITE (brittle silver ore) Ag_5SbS_4 rhomb. D: 6·2—6·3 H: $2^1/_2$ C: lead grey to iron black, often tarnished O: hydrotherm.

STIBNITE → antimony glance

STRONTIANITE $SrCO_3$ rhomb. D: 3·7 H: $3^1/_2$ C: colourless or coloured O: hydrotherm., sediment.

SULPHUR S rhomb. D: 2·1 H: 2 C: sulphur yellow O: volcan., sublim., sediment. Pls. 5, 58

SYLVANITE (graphic tellurium) $AuAgTe_4$ monocl. D: 8·0—8·3 H: $1^1/_2$—2 C: steely grey to silver white O: hydrotherm

SYLVINE KCl cub. D: 1·99 H: 2 C: colourless or coloured O: sal., sublim.

TALC (steatite) $Mg_3(OH)_2[Si_4O_{10}]$ monocl. D: 2·7—2·8 H: 1 C: green to

colourless, white, grey O: metam.

TANTALITE → columbite

TENORITE (melaconite) CuO monocl. D: 6 H: 3—4 C: black O: oxidat.

TETRADYMITE Bi_2Te_2S rhomb. D: 7·2—7·9 H: $1^1/_2$—2 C: light lead grey O: hydrotherm

TETRAHEDRITE (fahlerz) $(Cu_2,Ag_2,Fe,Zn,Hg)_3(Sb,As,Bi)_2S_6$ cub. D: 4·4—5·4 H: 3—4 C: steel grey O: hydrotherm. Pl. 52

THOMSONITE $NaCa_2[Al_2(Al,Si)Si_2O_{10}]_2 \cdot 5H_2O$ rhomb. D: 2·3—2·4 H: 5—$5^1/_2$ C: white, grey, yellowish, red O: volcan.

THORIANITE $(Th,U)O_2$ cub. D: 8—9·7 H: $5^1/_2$—$6^1/_2$ C: black O: pegmat., placer

THORITE (orangite) $ThSiO_4$ tetrag. D: 4·4—4·8 H: $4^1/_2$ C: orange, dark brown to black O: pegmat.

THULITE = pink-red zoisite

THURINGITE $(Fe,Al,Mg)_6[(OH)_8/(Al,Si)_4O_{10}]$ monocl. D: 3·15—3·19 H: $2^1/_2$ C: dark or light green, colourless O: sediment., metam.

TIGER'S EYE crocidolite altered to a golden-yellowish colour (due to limonite) and infiltrated with quartz. Pl. 138

TINCAL → borax

TOPAZ $Al_2[F_2/SiO_4]$ rhomb. D: 3·5—3·6 H: 8 C: colourless, yellow, blue, pink O: pneumat., placer Pls. 113, 132

TORBERNITE (copper uranite) $Cu[UO_2/PO_4]_2 \cdot 10H_2O$ tetrag. D: 3·2—3·3 H: 2—2$^1/_2$ C: green O: oxidat. Pl. 77

TOURMALINE (Na,Li, Ca) $(Mn,Mg,Fe,Al,Ti,Cr)_9$ $[(OH,F)_4/(BO_3)_3/Si_6O_{18}]$ trig. D: 3—3·25 H: 7 C: colourless, coloured or black O: pegmat., pneumat. Pls. 19, 36, 112

TREMOLITE $Ca_2(Mg,Fe)_5$ $[OH/Si_4O_{11}]_2$ monocl. D: 2·9—3·1 H: 5$^1/_2$ C: white, grey, greenish O: metam.

TRIDYMITE SiO_2 hex. D: 2·27 H: 6$^1/_2$—7 C: colourless, milky white or coloured O: volcan.

TROILITE FeS hex. D: 4·6 H: 4 C: brownish bronze O: meteor. Pl. 108

TURQUOISE $CuAl_6$ $[(OH)_8/(PO_4)_4] \cdot 4H_2O$

tricl. D: 2·6—2·8 H: 5—6 C: bluish green, sky blue O: sediment. Pls. 82, 144

ULEXITE (boronatrocalcite $CaNaB_5O_9 \cdot 8H_2O$ tricl. D: 2 H: 1 C: white O: sal.

UVAROVITE (calc-chrome garnet) $Ca_3Cr_2Si_3O_{12}$ cub. D: 3·4 H: 6$^1/_2$—7 C: dark emerald green O: metam.

VALENTINITE Sb_2O_3 rhomb. D: 5·6—5·8 H: 2—3 C: white, brownish O: oxidat.

VANADINITE Pb_5Cl $(VO_4)_3$ hex. D: 6·8—7·1 H: 3 C: yellowish, brownish, orange O: oxidat.

VESUVIANITE $Ca_{10}Al_4$ $(Mg,Fe)_2[(OH)_4/(SiO_4)_5(Si_2O_7)_2]$ tetrag. D: 3·27—3·45 H: 6$^1/_2$ C: green, brown, yellow, blackish brown O: metam., alp. joints

VIRIDINE manganiferous andalusite

VIVIANITE (Blue-iron earth) $Fe_3[PO_4] \cdot 8H_2O$ monocl. D: 2·6—2·7 H: 2 C: white, becoming blue O: oxidat., sediment.

WAD mixture of psilomelane and other manganese oxides

WAVELLITE $Al_3[(OH)_3/(PO_4)_2] \cdot 5H_2O$ rhomb. D: 2·3 H: 3$^1/_2$—4 C: colourless, grey, yellowish, greenish O: hydrotherm., oxidat., sediment. Pl. 70-1

WILLEMITE $Zn_2[SiO_4]$ trig. D: 4·1 H: 5$^1/_2$ C: colourless or coloured O: oxidat., metam.

WITHERITE $BaCO_3$ rhomb. D: 4·28 H: 3$^1/_2$ C: colourless, grey, yellowish O: hydrotherm.

WOLFRAMITE (Fe,Mn) WO_4 monocl. D: 7·14—7·54 H: 5—5$^1/_2$ C: dark brown to black O: pneumat.

WOLLASTONITE $CaSiO_3$ monocl. D: 2·8—2·9 H: 4$^1/_2$—5 C: white, grey, yellowish, brownish O: metam., volcan.

WULFENITE $Pb[MoO_4]$ tetrag. D: 6·7—6·9 H: 3 C: lemon yellow to orange O: oxidat. Pl. 81

WURTZITE ZnS hex. D: 4 H: 3$^1/_2$—4 C: light to dark brown O: hydrotherm. Pl. 51

ZEOLITES mineral group → chabazite, harmo-

tome, heulandite,
laumontite, natrolite,
phillipsite, thomsonite.

ZINCBLENDE → sphalerite

ZINCITE (red zinc oxide)
ZnO hex.
D: 5·4—5·7 H: 4^1/$_2$—5
C: blood red
O: metam.

ZINNWALDITE K(Fe,Li)$_3$
(F,OH)$_2$[AlSi$_3$O$_{10}$]
monocl. D: 2·9—3·1
H: 2^1/$_2$ C: silvery,
pink violet, brownish
O: pneumat.

ZIRCON ZrSiO$_4$ tetrag.
D: 3·9—4·8 H: 7^1/$_2$
C: brown, brownish,
red, grey, colourless,

blue, yellow
O: magmat., pegmat.,
metam., placer Pl. 135

ZOISITE Ca$_2$Al$_3$[O/OH/
SiO$_4$/Si$_2$O$_7$] rhomb.
D: 3·23—3·38 H: 6
C: grey, yellowish,
greenish O: metam.
Pl. 93